Computational Thermo-Fluid Dynamics of Aluminothermic Welding Process

Ravi Govindram Kewalramani

Computational Thermo-Fluid Dynamics of Aluminothermic Welding Process

Numerical Modelling of the Pouring and Solidification Stages

Springer Vieweg

Ravi Govindram Kewalramani
Professur für Technische Thermodynamik, Institut für Wärmetechnik und Thermodynamik
Fakultät für Maschinenbau, Verfahrens- und Energietechnik, TU Bergakademie Freiberg
Freiberg, Germany

Dissertation, angenommen von der Fakultät für Maschinenbau, Verfahrens- und Energietechnik der TU Bergakademie Freiberg, Freiberg, Deutschland, 2024

Doctoral Thesis, approved by the Faculty of Mechanical, Process and Energy Engineering of the TU Bergakademie Freiberg, Freiberg, Germany, 2024

Composition of the doctoral committee:
Reviewers: Prof. Dr.-Ing. Tobias M. Fieback (Doktorvater), TU Bergakademie Freiberg
Prof. Dr.-Ing. habil. Rüdiger Schwarze, TU Bergakademie Freiberg
Chairperson: Prof. Dr.-Ing. habil. Andreas Richter, TU Bergakademie Freiberg
Individual members: Prof. Dr. Subhashis Ray, TU Bergakademie Freiberg
Prof. Dr.-Ing. Gotthard Wolf, TU Bergakademie Freiberg
Dr.-Ing. IWE Jan Hantusch, Elektro-Thermit GmbH & Co. KG

ISBN 978-3-658-47534-5 ISBN 978-3-658-47535-2 (eBook)
https://doi.org/10.1007/978-3-658-47535-2

This Springer Vieweg imprint is published by the registered company Springer Fachmedien Wiesbaden GmbH, part of Springer Nature.
The registered company address is: Abraham-Lincoln-Str. 46, 65189 Wiesbaden, Germany

Acknowledgments

The research work reported in this doctoral thesis was carried out during my employment as a research associate at the Professorship for Technical Thermodynamics, Institute of Thermal Engineering, in the Faculty of Mechanical, Process and Energy Engineering, TU Bergakademie Freiberg. First and foremost, I would like to express my heartfelt gratitude to my thesis supervisor **Prof. Dr.-Ing. Tobias M. Fieback** for giving me the opportunity to work on this challenging numerical project and to be a part of his research group. In addition, I am thankful to **Prof. Dr.-Ing. habil. Rüdiger Schwarze** for kindly agreeing to serve as the second assessor.

I owe profound thanks to **Dr.-Ing. Ingo Riehl**, my daily supervisor, for his constant support and insightful discussions throughout my thesis. His open-door policy for both professional and personal discussions has been invaluable. His thoughtful feedback, constructive criticism and encouragement were instrumental in shaping both my academic journey and personal growth.

I extend my heartfelt gratitude to **Dr.-Ing. Jan Hantusch**, Senior Specialist Materials/Joining, Elektro-Thermit GmbH & Co. KG for bringing his expertise to the table and for his fruitful and insightful discussions throughout this research work. I also express my sincere gratitude to **Dr.-Ing. Jörge Keichel** and **Dr.-Ing. Matthias Wewel** from Elektro-Thermit GmbH & Co. KG, and **Dr.-Ing. Martin Niederkrüger**, Goldschmidt Holding GmbH, for their valuable suggestions and helpful comments.

Special thanks go to **Thomas Reinicke** for his indispensable support with the challenging experiments carried out at the Technology Innovation Center, particularly those related to achieving the melted rail profile post-pouring stage, which has been discussed in this work.

I greatly appreciate the assistance provided by **Sebastian Pose** in helping me understand the concepts of photogrammetry.

A huge shout-out to the amazing employees and former employees of the Institute of Thermal Engineering, with whom I've had the pleasure of working. Furthermore, Special thanks goes to **Aline Jünger, Doreen Bayer-Buhr, Dhaval Kulkarni, Florian Rau, Ingo Riehl, Kabriil Khajryan, Lukas Oppelt, Pitt Götze, Rhena Wulf, Sebastian Pose, Sebastian Weiss, Simon Eichinger, Thomas Grab und Thomas Storch** for creating an excellent working environment and fostering an open exchange of ideas.

A special note of thanks to **Dr.-Ing. Martin Heinrich** for his careful proofreading of Chap 3 and his insightful scientific suggestions. I also wish to express my gratitude to **Kabriil Khajryan** for proofreading the final draft of my thesis.

I would like to thank **Ms. Stefanie Meier** for her help and support with all the administrative work.

I am incredibly grateful to **Philipp Unglaub**e and his team (**Sebastian Conrad**, **Ulrich Herzig** and **Marco Köhler**) at the workshop for their unwavering support throughout my thesis. Their expertise and dedication were crucial in bringing my CAD models, concepts and design to life and also preparing the specimens for experiments in a timely manner. Moreover, I also thank **Almut Hegewald**, **Michael Wolf** and **Mirko Oehme** for offering their assistances when ever help was required.

I am immensely grateful to the Federal Ministry for Economic Affairs and Climate Action (BMWK), previously known as the Federal Ministry for Economic Affairs and Engery (BMWi), in Germany, for their generous financial support towards the MOTEVAS and PROENOS projects under project numbers 03ET1192B and 03EN2085B, respectively.

I also extend my appreciation to **Oliver Rebentrost** for his technical assistance and for providing useful tips on effectively utilising the HPC-Cluster resources. My appreciation also extends to the Faculty of Mathematics and Computer Science of Technische Universität Bergakademie Freiberg for the computing time on the HPC-Cluster, operated by the computing center (URZ) and funded by the German Research Foundation (DFG) under grant number 397252409.

On a personal note, I would like to express my heartfelt gratitude to my colleague-turned-friend, **Pitt Götze**, and his family for their warm hospitality during various celebrations/festivities. Their kindness and generosity made me feel truly welcomed and valued.

Finally, I am profoundly thankful to my family: my father, **Govindram T. Kewalramani**, my mother, **Leena G. Kewalramani**, and my sister, **Meenu G. Kewalramani**. Their unwavering love, encouragement and support have always been there for me, serving as a constant source of motivation and strength. I thank them from the bottom of my heart for everything they have done for me and for inspiring me to keep moving forward.

Abstract

The aluminothermic (AT) welding, commonly known as **Thermit** welding (TW), is an important process for joining and repairing rails due to its simplicity, robustness, portability and economic usage. It has been successfully used worldwide for over a century. It is, in principle, a casting process in which the molten metal from the crucible is poured into a preheated mould and allowed to solidify. For an efficient and high-quality weld, the knowledge of each stage involved during welding, such as preheating, thermite reaction, mould filling (pouring), and cooling (solidification) process, is significant.

This research aims to develop a numerical model to investigate the thermal pouring and cooling stages during the AT welding of rails. The multi-fluid system is modelled using the Volume of Fluid (VoF) method, while the enthalpy-porosity technique is employed for predicting the solid-liquid phase change. The $k - \varepsilon$ turbulence model is utilised to predict the turbulent flow field during the mould-filling process, which is also adapted to account for the solidification process. The entire model is developed and implemented in the Finite Volume Method (FVM) based open-source Computational Fluid Dynamics (CFD) software OpenFOAM. The model is validated against well-documented cases in the literature, demonstrating good agreement with reported results. To explore limitations of the proposed model, an in-situ experiment is designed and fabricated to study the melting of wax within enclosure in the presence of a gas phase. The numerical results show a good agreement with experiments and the deviations are discussed.

Due to the involvement of extremely high temperatures during the process and the opaque welding set-up, observing the flow field and phase change process inside the mould in real time is notoriously difficult. Experiments are designed and conceptualised to obtain the shape of the melted rails within the mould post-pouring stage to verify the numerical results of the thermal pouring stage. The experimentally acquired melted rails are digitalised using photogrammetry to capture intricate shape of the melted region post-pouring stage. Additional experiments are performed to record the temperature distribution in the rail during the AT welding process and the maximum width of the Fusion Zone (FZ) and Heat Affected Zone (HAZ). The procedure for the rapid welding process without preheating (RWOP) is followed for simplicity.

The developed numerical model is employed to simulate the thermal pouring stage. The influence of various pouring temperatures on the melting of rails post-pouring stage is investigated. The numerically obtained melted rails are compared with the experiments, and the observed discrepancies are discussed in detail. Moreover, the thermal pouring stage simulations are very computationally intensive and time consuming. Therefore, to investigate the solidification process during cooling stage assumptions are made to perform simulations on a simplified computational domain. The flow field during the solidification stage is considered to be laminar. Additionally, the influence of various molten metal temperatures and thermophysical properties on the FZ are investigated.

Zusammenfassung

Das Aluminothermische (AT) Schweißen, auch bekannt als **Thermit**-Schweißen, ist ein jahrhunderte-alter, weltweit verwendeter Prozess zum Schweißen und Reparieren von Schienen, begründet durch seine Einfachheit, Robustheit, Portabilität und Wirtschaftlichkeit. Es ist im Prinzip ein Gießverfahren, bei dem das geschmolzene Metall aus dem Tiegel in eine vorgewärmte Form strömt und dort abkühlt. Für eine effiziente und qualitativ hochwertige Schweißung ist das Wissen über die einzelnen Phasen des AT-Schweißens (Vorwärmung, Thermit-Reaktion, Formfüllung, Abkühlung und Erstarrung) von entscheidender Bedeutung.

Das Ziel dieser Forschung ist die Entwicklung eines numerischen Modells zur Untersuchung der Gießphase und der Abkühlung/Erstarrung beim AT-Schweißen von Schienen. Das Multi-Fluid-Sys-tem wird mithilfe der Volume-of-Fluide (VoF) Methode modelliert, während die Enthalpie-Porositäts-Methode für die Modellierung des fest/flüssig Phasenwechsels verwendet wird. Das $k - \varepsilon$ Turbulenzmodell wird genutzt, um das turbulente Strömungsfeld während des Formfüllprozesses zu beschreiben. Es wird modifiziert, um den Erstarrungsprozess zu berücksichtigen. Das gesamte Mehrphasenmodell wird in der Finite-Volumen-Methode (FVM) basierten quell-offenen Software für numerische Strömungsmechanik (CFD) OpenFOAM implementiert. Das Modell wird anhand gut dokumentierter Fälle aus der Literatur validiert und zeigt eine gute Übereinstimmung mit den berichteten Ergebnissen. Zur Erkundung der Grenzen des vorgeschlagenen Modells wird ein In-situ-Experiment entworfen und hergestellt, um das Schmelzen von Wachs in einem Gehäuse in Gegenwart einer Gasphase zu untersuchen. Die numerischen Ergebnisse zeigen eine gute Übereinstimmung mit den Experimenten, und die beobachteten Abweichungen werden diskutiert.

Aufgrund der extrem hohen Temperaturen während des Prozesses und der undurchsichtigen Schweiß-anordnung ist es schwierig, das Strömungsfeld und den Phasenwechselprozess innerhalb der Form zu beobachten. Experimente werden entworfen und konzipiert, um angeschmolzene Schienen innerhalb der Form unmittelbar nach der Gießphase zu erhalten und damit die numerischen Ergebnisse der thermischen Gießphase zu überprüfen. Die experimentell erfassten angeschmolzenen Schienen werden mithilfe von Photogrammetrietechniken digitalisiert, um die komplizierte Form der angeschmolzenen Schienen nach der Gießphase analysieren zu können. Zusätzliche Experimente werden durchgeführt, um die Temperaturverteilung in der Schiene während des AT-Schweißprozesses und die maximale Breite der Schmelzzone (SZ) und der Wärmeeinflusszone (WEZ) zu erfassen. Das Schnell-schweißverfahren ohne Vorwärmung (SoV) wird aus Gründen der Einfachheit verwendet.

Das numerische Modell wird zur Simulation der thermischen Gießphase eingesetzt. Der Einfluss verschiedener Gießtemperaturen auf die Aufschmelzung der Schienen nach dem Ende der Gießphase wird untersucht. Die numerisch erhaltenen angeschmolzenen Schienen werden mit den Experimenten verglichen und die beobachteten Abweichungen werden ausführlich erörtert. Die Simulationen der thermischen Gießphase sind sehr rechenintensiv und zeitaufwändig. Daher werden Annahmen getroffen, um den Erstarrungsprozess während der Abkühlphase auf einem vereinfachten Rechengebiet zu simulieren. Das Strömungsfeld während der Erstarrungsphase wird als laminar betrachtet. Zusätzlich werden der Einfluss verschiedener Temperaturen der Schmelze und thermophysikalischer Eigenschaften auf die SZ untersucht.

Contents

1 Introduction 1

2 State of the Art 5
2.1 Aluminothermic Welding Process 5
2.2 Literature Survey 7
2.3 Contribution of this PhD Thesis 15

3 Numerical Modelling 19
3.1 Governing Equations 19
3.2 Volume of Fluid Method 22
3.3 Solid-liquid Phase Change Model 26
3.4 Multiphase Turbulence Modelling 29
3.5 Numerical Methods 32
3.6 OpenFOAM 47

4 Validation 49
4.1 One-dimensional two-phase Stefan Problem 49
4.1.1 Problem Description 49
4.1.2 Results and Discussions 53
4.2 Melting of Pure Gallium in a Rectangular Enclosure 55
4.2.1 Problem Description 55
4.2.2 Results and Discussions 56
4.3 Free Surface Thermo-capillary Driven Flow with Phase Change 61
4.3.1 Problem Description 61

4.3.2 Results and Discussions 63
4.4 Melting of Paraffin Wax in the Presence of Gas Phase 64
4.4.1 Experimental Set-up 64
4.4.2 Simulation set-up 66
4.4.3 Results and discussions 68

5 Exploring Aluminothermic Welding of Rails Through Experimental Investigations 73
5.1 Experimental Set-up and Procedure 74
5.2 Photogrammetric 3D Digitalisation of the Melted Rails 76
5.3 Shape of the Melted Rails 78
5.4 Temperature Measurements 80
5.5 Fusion and Heat Affected Zone 83

6 Simulation of the Thermal Pouring Stage 87
6.1 Simulation Set-up 87
6.2 Results and Discussions 92
6.2.1 Mesh Sensitivity Analysis 92
6.2.2 Computational Speed-up 93
6.2.3 Influence of the Pouring Temperature 95
6.2.4 Influence of the Draining Process 107

7 Simulation of the Solidification Process 111
7.1 Simulation Set-up 111
7.2 Results and Discussions 115
7.2.1 Mesh Sensitivity Analysis 115
7.2.2 Melting/Solidification 117
7.2.3 Flow Field Inside the Mould 118
7.2.4 Influence of the Initial Molten Metal Temperature 119
7.2.5 Influence of the Thermophysical Properties 126
7.2.6 Rapid Welding Process with Short Preheating 131

8 Closure 139

List of Publications 145

Bibliography 147

Nomenclature

Abbreviations

AT	Aluminothermic
BD	Blended Differencing
CCT	Circular Coded Target
CDS	Central Differencing Scheme
CFD	Computational Fluid Dynamics
CFL	Courant-Friedrichs-Lewy
CV	Control Volume
CWR	Continuously Welded Rails
DILU	Diagonal-based Incomplete Lower-Upper
DSLR	Digital Single Lens Reflex
FEM	Finite Element Method
FVM	Finite Volume Method
FZ	Fusion Zone
GAMG	Geometric agglomerated Algebraic Multigrid
GPL	General Public Licence
HAZ	Heat-Affected Zone
HPC	High Performance Computing
HRS	High Resolution Schemes
LED	Light Emitting Diode
LUDS	Linear Upwind Differencing Scheme
MULES	Multidimensional Universal Limiter for Explicit Solutions
NVA	Normalised Variable Approach

PBiCGStab	Preconditioned Bi-Conjugate Gradient Stabilised
PCG	Preconditioned Conjugate Gradient
PCM	Phase Change Material
PDE	Partial Differential Equations
RANS	Reynolds-Averaged Navier-Stokes
RT	Rubitherm®
RWOP	Rapid Welding process without Preheating
RWSP	Rapid Welding process with Short Preheating
SkV	Schnellschweißverfahren mit kurzer Vorwärmung (**German**)
SoV	Schnellschweißverfahren ohne Vorwärmung (**German**)
SRC	Sinlge-use Reaction Crucible
TVD	Total Variation Diminishing
TW	Thermit welding
UDS2	Second-order Upwind Differencing Scheme
UDS	Upwind Differencing Scheme
VoF	Volume of Fluid
UIC	Union Internationale des Chemins de fer (**French**)

Greek Symbols

α	Volume fraction (–)
β	Thermal expansion coefficient (K^{-1})
δ	Kronecker delta function (–)
ϵ	Computational constant (–)
η	Eigen value (–)
Γ	Diffusivity (–)
γ	Melt volume fraction (solid/liquid) (–)
κ	Thermal conductivity (W $m^{-1}K^{-1}$)
ϕ	generic material property (–)
τ	Viscous stress tensor (kg $m^{-1}s^{-1}$)
φ	generic variable (–)
ϱ	Density (kg m^{-3})
ε	Turbulent dissipation (m^2s^{-3})
μ	Dynamic viscosity (kg $m^{-1}s^{-1}$)
ν	Kinematic viscosity (m^2s^{-1})

Latin Symbols

a	Thermal diffusivity (m^2s^{-1})
C	Darcy constant (kg $m^{-3}s^{-1}$)
c_p	Specific heat capacity (J kg $m^{-1}K^{-1}$)
g	Gravitational constant (ms^{-2})
H	Height (m)
h	Enthalpy (m^2s^{-2})
h_c	Convective heat transfer coeff. (W $m^{-2}K^{-1}$)
$\mathbf{I}$	Unit tensor (–)
k	Turbulent kinetic energy (m^2s^{-2})
$\mathcal{L}$	Latent heat of fusion ($kJkg^{-1}$)
L	Length (m)
N	Neighbour cell (–)
P	Owner cell (–)
p	Pressure (kg $m^{-1}s^{-2}$)
S	Source term (–)
S_f	Surface area (m^2)
S_{ij}	Strain rate tensor (s^{-1})
T	Temperature (K)
t	Time (s)
u	Velocity vector (ms^{-1})
V	Volume (m^3)
W	Width (m)

Subscripts

1	Fluid 1
2	Fluid 2
MM	Molten Metal
R	Rail
b	Body force
c	Cold wall
eff	effective
f	cell face
h	Hot wall
i	fluid index

ini	Initial
l	Liquid
mom	Momentum
ref	Reference
s	Solid
t	turbulent
x	x-direction
y	y-direction
z	z-direction

Superscripts

$()^T$	Transpose of a matrix or vector
$()^R$	Reynolds

List of Figures

Figure 2.1	Schematic of the AT welding set-up: Full assembly (left), sectional view (right)	6
Figure 3.1	A schematic representation of the volume fraction field in the VoF method, depicting the interface between liquid Thermit-steel (grey) and air (white)	23
Figure 3.2	Schematic illustration of control volume and finite volume discretisation parameters, adapted from [41, 53]	33
Figure 3.3	Schematic representation of the vectors d and S_f on a non-orthogonal mesh	35
Figure 3.4	Illustration of UDS on 1D equidistant grid	37
Figure 3.5	Illustration of CDS on 1D equidistant grid	38
Figure 3.6	Illustration of LUDS on 1D equidistant grid	38
Figure 3.7	Graphical representation of the developed numerical model for predicting the dynamics of 3D multi-fluid, multiphase heat transfer and solid-liquid phase change throughout the mould filling and cooling stages of the AT welding process	46
Figure 4.1	Illustration of the 1D two-phase Stefan problem	51
Figure 4.2	Locations of the solid-liquid interface front for three different combinations of Stefan numbers in solid and liquid phases: Analytical versus Simulation	54

Figure 4.3 Temperature distribution in the slab at three different time instances for the combination 2: Analytical versus Simulation ... 55
Figure 4.4 Numerical set-up of gallium melting test case ... 56
Figure 4.5 Comparison of the solid-liquid interface fronts from experiments [73] and 2D simulations for various grid configurations at different times ... 57
Figure 4.6 Comparison of the solid-liquid interface fronts from experiments [73] and 2D simulations for various numerical discretisation schemes at different times ... 58
Figure 4.7 Comparison of the solid-liquid interface fronts from 2D and 3D simulations with experiments [73] at different times ... 59
Figure 4.8 Comparison of the solid-liquid interface fronts from 3D simulation with experiments [73] and the 3D numerical results reported in the literature [24, 25, 76, 79] at different times ... 60
Figure 4.9 Numerical set-up of the rectangular domain filled with bismuth metal and argon gas ... 61
Figure 4.10 Numerical prediction of the velocity flow field, bismuth-argon interface (blue) and solid/liquid interface (red) at steady state ... 63
Figure 4.11 Comparison of the solid-liquid interface front at steady state with numerical results of Saldi [78] and Tan et al. [84] ... 64
Figure 4.12 Schematic representation of the experimental set-up ... 65
Figure 4.13 Numerical set-up for simulating the melting of wax in the presence of gas phase ... 67
Figure 4.14 Comparison of the solid-liquid interface fronts at time $t = 25$ min for three distinct grid configurations ... 69
Figure 4.15 Qualitative analysis of the solid-liquid interface fronts: Experiment versus sim-ulation ... 70
Figure 4.16 Quantitative analysis of the solid-liquid interface fronts: Experiments versus simulation ... 71
Figure 5.1 Experimental set-up to obtain the shape of the melted rails post-pouring stage ... 75
Figure 5.2 The shape of the melted rails obtained immediately after the pouring stage ... 75

Figure 5.3 A turntable set-up for capturing a series of overlapping photographs for the three dimensional reconstruction of the rail (Experiment: left rail) ... 76
Figure 5.4 Workflow for image processing and 3D reconstruction in Agisoft Metashape [95] ... 77
Figure 5.5 Photogrammetry of the experimentally obtained melted rails within the mould—**Note:** Herein, the photographs are acquired from four different positions ... 78
Figure 5.6 Qualitative analysis of the shape of the melted rails within the mould post-pouring stage: Experiment and Photogrammetry ... 79
Figure 5.7 Width of the gap between the melted rails post-pouring stage ... 80
Figure 5.8 Measurement locations of thermocouples: schematic representation (top, adapted from [9]), experiment (bottom) ... 81
Figure 5.9 Temperature distribution in the rail cross head (H1, H2), web (W1) and base (F1) regions during mould filling and cooling stages ... 82
Figure 5.10 Temperature distribution in the rail at base (F2, F3) regions during mould filling and cooling stages ... 83
Figure 5.11 Metallographic analysis of the longitudinal section across the rail joint [97] ... 84
Figure 5.12 Experimentally obtained maximum width of FZ and HAZ ... 84
Figure 6.1 Mesh generation strategy implemented for simulating pouring process ... 88
Figure 6.2 Initialisation of field parameters, domain boundaries and boundary conditions [99] ... 89
Figure 6.3 Comparison of the flow filed inside the mould at time $t = 0.35$ s for different values of Δx in mm ... 93
Figure 6.4 Computational speed-up obtained for the multiphase mould filling simulation of AT welding process on the HPC-Cluster of the TU Bergakademie Freiberg ... 94
Figure 6.5 Velocity vectors in the liquid Thermit-steel and the temperature distribution depicting the melting of the rail during the pouring stage for the pouring temperature $T_P = 2073$ K ... 96

Figure 6.6 Temperature field of the liquid Thermit-steel during the pouring stage for the pouring temperature $T_P = 2073$ K 97
Figure 6.7 Velocity vectors in the liquid Thermit-steel and the temperature distribution depicting the melting off of the rail during the pouring stage for the pouring temperature $T_P = 2273$ K 98
Figure 6.8 Temperature field of the liquid Thermit-steel during the pouring stage for the pouring temperature $T_P = 2273$ K 99
Figure 6.9 Velocity vectors in the liquid Thermit-steel and the temperature distribution depicting the melting off of the rail during the pouring stage for the pouring temperature $T_P = 2473$ K 100
Figure 6.10 Temperature field of the liquid Thermit-steel during the pouring stage for the pouring temperature $T_P = 2473$ K 101
Figure 6.11 Velocity vectors in the liquid Thermit-steel and the temperature distribution depicting the melting off of the rail during the pouring stage for the pouring temperature $T_P = 2673$ K 102
Figure 6.12 Temperature field of the liquid Thermit-steel during the pouring stage for the pouring temperature $T_P = 2673$ K 104
Figure 6.13 Peak temperatures at the weld center after the end of pouring stage (i.e., at t =6.50 s) for various pouring temperatures investigated in this study 105
Figure 6.14 Qualitative analysis of the shape of the melted rail profiles within the mould post-pouring stage: Experiment (top left), photogrammetry (top right) and numerically obtained melted rail profile at various pouring temperatures—**Note:** The casting mould (transparent green) is shown only for better understanding 106
Figure 6.15 Comparison of the dimensions of the melted rails within the mould post-pouring stage from experiments and simulation for different pouring temperatures 107

Figure 6.16 Instantaneous velocity vector field in the liquid Thermit-steel and the melted rail within the mould during the draining process at $t = 0.15$ s for case with a pouring temperature $T_P = 2473$ K 108
Figure 6.17 Comparison of the width of the weld gap between the melted rails before (at $t = 5$ s) and after draining the liquid Thermit-steel in the simulation at a pouring temperature $T_P =$ 2473 K along with experimental data 109
Figure 7.1 Initialisation of fields: melt fraction, volume fraction, velocity and temperature field at time $t = 0$ s [97] 112
Figure 7.2 An exploded view of all the boundaries of the computational domain [97] 113
Figure 7.3 Position of the solid-liquid interface front for different cell sizes at time $t = 10$ s 116
Figure 7.4 The solid-liquid interface front for the initial molten metal temperature of 2473 K at $t = 2$ s, 10 s and 40 s ... 117
Figure 7.5 Flow field inside the mould during the initial stages of cooling: velocity vectors (coloured), mould (transparent green), 3D solid liquid interface profile and parent rail (solid grey) [97] 118
Figure 7.6 Temporal and spatial evolution of the solid-liquid interface front at the initial molten metal temperature of 2073 K 120
Figure 7.7 Temporal and spatial evolution of the solid-liquid interface front at the initial molten metal temperature of 2273 K 121
Figure 7.8 Temporal and spatial evolution of the solid-liquid interface front at the initial molten metal temperature of 2473 K 123
Figure 7.9 Temporal and spatial evolution of the solid-liquid interface front at the initial molten metal temperature of 2673 K 124
Figure 7.10 Comparison of the maximum width of the FZ (a) and HAZ (b) between experiments and simulations for the molten metal temperatures 2473 K (blue) and 2673 K (red) 125

Figure 7.11 Comparison of the temperature distribution in the rail at head (H1), web (W1) and base (F1, F3) regions between experiment and simulations for the molten metal temperatures 2473 K (blue) and 2673 K (red) 126
Figure 7.12 Comparison of the maximum width of the FZ on the symmetry plane along the x-axis between: Experiments (symbols) versus reference simulation (black solid line with symbol) with various density (solid, liquid) values in kg m^{-3} (coloured solid lines with symbols) 127
Figure 7.13 Comparison of the maximum width of the FZ on the symmetry plane along the x-axis between: Experiments (symbols) versus reference simulation (black solid line with symbol) with an alternative dynamic viscosity in kg m^{-1} s^{-1} (coloured solid line with symbols) 128
Figure 7.14 Comparison of the maximum width of the FZ on the symmetry plane along the x-axis between: Experiments (symbols) versus reference simulation (black solid line with symbol) with various specific heat capacity values in J kg^{-1} K^{-1} (coloured solid lines with symbols) 129
Figure 7.15 Comparison of the maximum width of the FZ on the symmetry plane along the x-axis between: Experiments (symbols) versus reference simulation (black solid line with symbol) with various thermal conductivity values in W m^{-1} K^{-1} (coloured solid lines with symbols) 130
Figure 7.16 Temporal evolution of the temperature field at different locations in the rail during preheating and tapping period: Experiment vs numerical prediction 132
Figure 7.17 The molten metal inside the mould is initialised with the temperature of 2473 K and the temperature distribution in the rail after the preheating stage (including tapping period) at t =190 s 133

Figure 7.18 Temporal and spatial evolution of the solid-liquid interface front for the initial molten metal temperature of 2473 K with approx 2.5 min of preheating followed by approx 30 s of tapping period 134
Figure 7.19 Comparison of the location of the solid-liquid interface front on the symmetry plane along the x-axis between RWOP and RWSP processes for the initial molten metal temperature of 2473 K at $t = 6$ s, 30 s, 40 s 135
Figure 7.20 Comparing the maximum width of the FZ and HAZ in the RWSP process: Experimental data versus simulation results 136

List of Tables

Table 2.1 Overview of the literature review discussed in this chapter. The description of the abbreviations are as follows: C—Conduction, CC—Conduction and Convection, I—Isothermal, HFP—Heat Flux Profile, Sim—Simulation 16
Table 4.1 Material properties of steel in solid and liquid phases 52
Table 4.2 Computed eigenvalues and Stefan numbers for different temperatures imposed on the hot and cold walls 52
Table 4.3 Quantitative analysis of the locations of the interface front for 3 different combinations of Stefan numbers: Analytical versus Simulation – **Note:** A, S and RE denote Analytical, Simulation and Relative Error, respectively 54
Table 4.4 Material properties of gallium 56
Table 4.5 Material properties of bismuth, argon gas and the relevant dimensionless numbers 62
Table 4.6 Boundary conditions for simulating the melting of wax in the presense of gas phase 67
Table 4.7 Material properties of RT31 68
Table 5.1 Measurement locations of thermocouples 82
Table 5.2 Measurements of the FZ (A-E) and HAZ (F-H) 85
Table 6.1 Material properties of the steel [9, 13, 101] 90
Table 6.2 Regression coefficients of the alloying elements; published in [97] 90

Table 6.3 Contact angle θ values of liquid Thermit-steel with crucible, mould, sand plug and rails used for the simulations . . . 91
Table 6.4 Numerical discretisation schemes . . . 91
Table 6.5 Total size of the computational domain for each Δx investigated in this study . . . 92
Table 7.1 Material properties employed in the simulations for liquid Thermit-steel, rail steel and slag [9, 13, 16, 101] . . . 114
Table 7.2 Numerical discretisation schemes . . . 115
Table 7.3 Size of the computational domain . . . 116
Table 7.4 Measurements of the FZ (A–E) and HAZ (F–H): Comparison between experimental data and simulation results for different molten metal temperatures (refer to Figure 5.11); published in [97]—**Note:** The grey colour in the table indicates melting below the required 3 mm limit . . . 125
Table 7.5 Comparing the measurements of the maximum width of the FZ (A–Z in Figure 5.11) for all simulated variants . . . 131

Introduction 1

Motivation

Following the approval of the Paris Agreement [1] in 2016, which aims to limit the global average temperature increase to well below 2 °C, the European Climate Law (ECL) [2] was put into action, in July 2021, declaring the EU's commitment to being climate neutral by the middle of this century. To work towards this objective, the law is set to cut down on greenhouse gas emissions by at least 55% compared to the levels in 1990 by the year 2030.[1] For instance, according to data presented in the Climate Action Plan 2050 (which signifies Germany's commitment to achieving climate neutrality) [3], indicates that carbon dioxide (CO_2) emissions within the transport sector were approximately 163 million tonnes in 1990. Remarkably, this level has remained relatively consistent in 2014, at around 160 million tonnes. With the enforcement of the ECL and alignment with the Paris Agreement, a projected reduction in CO_2 emissions to 95–98 million tonnes is anticipated, reflecting an overall reduction of about 42–40% compared to 1990.

In the global campaign against climate change, the European Commission is committed to invest in high-speed rails as part of its Sustainable and Smart Mobility Strategy. The aim is to increase the number of high-speed rails on European tracks by a factor of two by 2030 and by a factor of three by 2050 [4]. In tandem with this, the German government is actively developing a rail transport concept for 2030/2050, seeking to maximise the potential for road traffic to shift to railways [3]. Among other factors, two crucial elements may contribute to the reduction of

---

[1] The main goal is to achieve a reduction of 55% across all sectors, encompassing energy, building, transport, etc. Nevertheless, the specific contributions to the reduction may vary at an individual sector level.

R. G. Kewalramani, *Computational Thermo-Fluid Dynamics of Aluminothermic Welding Process*, https://doi.org/10.1007/978-3-658-47535-2_1

CO_2 emissions. Firstly, the electrification of railroads is a key focus to use clean energy and reduce the carbon footprint. According to Pro-Rail Alliance (German: *Allianz pro Schiene*), only 61% of the German rail network is electrified, with a target to increase this to 75% by 2030 [5]. Secondly, transporting cargo via rail is recognised as an effective strategy for reducing greenhouse gas emissions. However, to achieve this goal, there is a need to shift more freight from road transportation to rail. For instance, based on data available from 2017, the rail's share of global freight transport in Germany is less than 17%, while a substantial 72% is reliant on road transport [6].

Anticipating a surge in both passengers and heavy haul freight trains in the near future, rails (railway tracks) are expected to endure increased mechanical stresses and heavier axle loads in the near future. Consequently, with the rising rail traffic, there will be a growing demand to prioritise the repair and maintenance of rails, while simultaneously minimising disruptions to rail operations.

The historical auditory signature of bolted rail joints, characterised by the rhythmic "clickity-clack", encapsulates an era where the history of train travel is resonated with by each metallic beat. However, this auditory signature is not just a nostalgic melody, but it also serves as a tangible indicator of challenges within bolted joints on railway tracks: Issues such as the need for frequent maintenance, susceptibility to wear and tear, reduced speed and efficiency, among others, have prompted a paradigm shift towards Continuously Welded Rails (CWR). This transition marks a crucial advancement in the railway industry. The advantages of CWR in railroad tracks are substantial, offering a smoother ride experience, enabling higher speeds, and less maintenance compared to conventional bolted joints [7].

In the railway industry, the production of CWR involves various welding methods, such as gas pressure welding, enclosed arc welding, flash butt welding and well-established aluminothermic (AT) welding, commonly known as Thermit welding. The AT welding process is highly regarded for its simplicity, robustness, portability and cost-effectiveness in rail joining and repairing applications and has been successfully used globally for over a century.

The AT welding process essentially resembles a casting technique in which the molten metal from a crucible is poured into a preheated mould and allowed to solidify. However, this similarity also means that the process is susceptible to various casting defects, including shrinkage defects, metallurgical defects, pouring metal defects, etc., ultimately compromising the quality of the weld. Therefore, in order to achieve an efficient and high-quality weld, the knowledge of thermal conditions during each stage of the process is significant. However, conducting experiments alone proves challenging due to the extreme temperatures involved during the process and the opaque nature of the welding set-up.

In this study, the primary focus is to gain a deeper understanding of the thermal pouring and cooling stages during the AT welding process by means of numerical investigations. To achieve this goal a comprehensive numerical model is developed and rigorously validated through benchmark cases and in-situ experiments.

Outline

The objectives and the structure of the current work are as follows:

Chapter 2 briefly describes the AT welding process which is used for joining and repairing rails. Additionally, it presents a comprehensive review of the existing literature that highlights the previously conducted numerical investigations related to the AT welding process. Finally, the chapter outlines the contributions of the research work reported in this PhD thesis.

Chapter 3 introduces the open-source CFD library used in this study, i.e., OpenFOAM. Moreover, the governing equations required to predict the flow field, heat transfer and solidification/melting during the AT welding process are discussed in detail.

Chapter 4 systematically validates the numerical model based on well-documented benchmark cases from the literature. First, the one-dimension Stefan problem is numerically investigated and compared with the analytical solution. Next, the melting of gallium metal in an enclosure is studied, followed by the discussion of the influence of simulation parameters on the temporal and spatial evolution of the melting front. Then, the solidification of liquid bismuth with a free surface is performed to demonstrate the robustness of the developed model. Finally, in-situ experiments are conducted to study the melting of wax (Rubitherm 31) within an enclosure.

Chapter 5 presents validation experiments of the AT welding process in laboratory conditions to obtain the temporal and spatial evolution of the temperature in the rail and the maximum width of the fusion and heat-affected zone. The experiments were conceptualised to obtain the shape of the melted rails within the mould immediately after the pouring stage. Furthermore, a digital model of the experimentally obtained melted rails is obtained using photogrammetry. The experimental trials were conducted at the Technology Innovation Center of Goldschmidt Holding GmbH. The experiments were based on SoV-process (German: Schnellschweißverfahren ohne Vorwärmung), i,e., rapid welding process without preheating (RWOP).

Chapter 6 deals with the simulations of the thermal mould-filling stage, including solid-liquid phase change. The optimum simulation settings are identified to simulate this computationally intensive process. The influence of the the initial temperature of the liquid Thermit-steel on the flow field and the shape of the 3D melted rail profiles are investigated.

Chapter 7 serves to validate the numerical simulations of the cooling stage of the AT welding process. The numerically obtained temperature distribution in the rail and the maximum width of the fusion and heat-affected zone are compared with experiments. In addition, a parameter study is performed to investigate the influence of various parameters on the temporal evolution of the solid-liquid interface front and the final weld.

Chapter 8 summarises the main findings, limitations and conclusions of the present work. Additionally, it offers some recommendations for extending and improving the developed numerical model.

State of the Art

2

2.1 Aluminothermic Welding Process

In the AT welding process, the Thermit-mixture[1] (comprising finely powdered iron(III) oxide: Fe_2O_3, aluminium: Al and other alloying elements to improve the mechanical properties of the weld metal) undergoes an exothermic redox reaction. In this reaction aluminium acts as the reducing agent, leading to the reduction of iron oxide to iron metal Fe at high temperatures:

$$Fe_2O_3 + 2Al- > 2Fe + Al_2O_3, \Delta H_R^0 = -850 \text{kJ mol}^{-1}, \tag{2.1}$$

where ΔH_R^0 is the enthalpy of reaction. The heat generated during this reaction process is sufficient to produce low-viscosity molten steel required for the complete welding of the rail track and slag (alumina: Al_2O_3) as a byproduct. As aluminium is used as a reducing agent, this reaction is called an AT reaction. The most commonly known AT reactions can be found in the works of Jha [8], Tuchkova [9] and Wegrzyn and Maxur [10]. Figure 2.1 shows a typical Thermit SkV-Elite L25 mould assembly for a 60E1 (formerly known as UIC60) [11] rail profile, here SkV (German: *Schnellschweißverfahren mit kurzer Vorwärmung*) means rapid welding process with short preheating (RWSP). The AT welding of rails comprises of the following stages:

1. **Initial preparation:** The rail ends are cut and cleaned to maintain a standard weld gap of 25 mm. The rails are then slightly elevated to account for shrinkage during solidification. Next, a two-piece mould is mounted on either side of the

[1] For simplicity, the use of "Thermit" will be favoured over the registered trademark "Thermit®"

R. G. Kewalramani, *Computational Thermo-Fluid Dynamics of Aluminothermic Welding Process*, https://doi.org/10.1007/978-3-658-47535-2_2

rail and the assembly is sealed with luting sand to prevent any leakage of molten metal during the pouring process. Finally, the mould is packed with mould shoe to stabilise and clamp the casting assembly on the rail.

2. **Preheating:** The welding assembly is now preheated for 2.5 min or longer depending upon the type of welding with a Smartweld jet [12], which is designed especially for this purpose.
3. **Aluminothermic reaction:** Once the preheating stage is complete, the sand plug is inserted into the mould, followed by positioning the crucible over it. The Thermit-mixture is then ignited, triggering an exothermic redox reaction followed by the separation of steel and slag within the crucible.
4. **Pouring (crucible discharge and mould filling):** The discharge of the molten metal from the reaction crucible into the mould.
5. **Cooling/Solidification:** The molten metal in the mould is now allowed to cool and solidify for a certain period before the complete setup is disassembled.
6. **Final finishing:** The excess weld material is removed and the uneven areas on the surface of the rail head are smoothed out through grinding.

Thus, the final quality of the weld depends on each of the process steps (i.e., from steps 1 through 6) and the thermal conditions involved during welding.

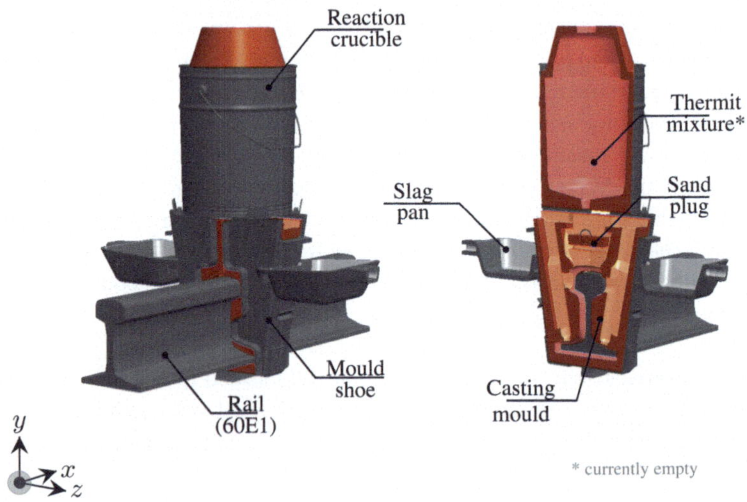

Figure 2.1 Schematic of the AT welding set-up: Full assembly (left), sectional view (right)

As previously mentioned, this study involves the development of a numerical model to predict the flow field and the solid-liquid phase change phenomena during the thermal pouring and cooling stages during AT welding process. The subsequent section presents a comprehensive overview of the existing numerical models/investigations, specifically addressing the AT welding process.

2.2 Literature Survey

In 1989, Jha [8] developed a finite element method (FEM) based heat transfer model to predict the peak temperatures, heating and cooling temperature profiles in the weld metal, heat affected zone (HAZ) and parent rail during Thermit[2] welding (TW) process. The purpose of the model was to identify an appropriate post-weld heat treatment that would improve the mechanical properties of the weld without negatively affecting the micro-structure in HAZ. The study included experiments at the Association of American Rail Road, Technical Center with varying weld gaps using a standard Thermit-mixture available at the time. During the welding process, temperatures were measured at various locations in the rails. Due to its two-fold symmetry only a quarter of the rail geometry was considered for numerical investigation. The preheating simulations were carried out by specifying different values of heat flux at rail head, web and base region. The pouring stage was assumed to be instantaneous and different liquid metal temperature values were used to initialise the rail head, web and rail base region. Moreover, the convection in the molten metal was considered by artificially increasing the thermal conductivity by 20 %. The temperature measurement obtained from the simulations showed good agreement with those from the experiments. Subsequently, mechanical properties, macro-and micro-structure formation and heat effected zone for each weld were also analysed after the joining process. The analysis revealed that the mechanical properties (ductility and impact properties) of the weld metal was inferior to the parent rails. Therefore, the specimens were subjected to normalising process to improve the mechanical properties of the weld. During the normalising process, the specimens prepared for different tests (tensile, charpy, etc.) were heated to a minimum austenising temperature and allowing it to cool down to room temperature. This heating and slow cooling alters the micro-structure formation in the weld, resulting in improved ductility. The normalising treatment showed a significant improvement in the percentage elongation of the weld, increase in yield and ultimate tensile strength. However, the challenging part was to realise the normalising process

[2] To maintain consistency the word "thermite" used in the literature is replaced with "Thermit"

on the rail profile in the field using a oxy/propane tourch. To address this issue, Jha further developed the validated heat transfer model to investigate the effect of oxy/propane tourch on the normalisation process of the weld metal, HAZ and parent rail. He then performed numerical experiments with two different heat flux values (high and low) to investigate the temperature distribution in the welded rail section during the normalising process. His results predicted, that the high heat flux values resulted in a steep temperature gradients in both longitudinal and transverse direction, whereas low heat flux values gave more uniform temperature distribution in the transverse direction and steep temperature gradients in longitudinal direction. Thus, the low heat flux values were found to be more optimal to obtain the required temperature distribution for the post-weld heat treatment. This model is one of the first numerical approach to predict the heat transfer process during the TW of rails.

Chen et al. [13–15] developed an FEM-based heat transfer model to investigate the heat transfer process during the TW process. The computational domain investigated in their study was based on Orgo-Thermit SkV standard TW kits. Only one-fourth of the computational domain was considered due to symmetry with longitudinal and transverse central planes. The authors conducted simulations using FIDAP, a commercial FEM package for CFD. The mathematical modelling was exclusively based on heat conduction, and efforts were made to include other forms of heat transfer through initial or boundary conditions. In addition, laboratory welds[3] were performed to validate their numerical model. However, the authors adopted a hybrid approach to tackle this complex problem numerically. The experimental data were used to validate the numerical results and as input data for the preheating simulations and initial conditions for the solidification simulations. For the preheating simulations, the heat-flux profiles were first estimated from the temperature measurements obtained after each laboratory weld. Then, the experimentally estimated heat-flux profile was implemented as a boundary condition on the rail end (inside the mould), and the simulations were performed. Next, the numerically obtained rail temperature distribution at various positions was compared with the available experimental data. If the numerical results agreed well with the corresponding experimentally obtained temperature measurements, then the numerically obtained temperature distribution in the rail was used for the solidification simulations. Otherwise, the heat flux was adjusted until a good agreement with experiments was observed. For the solidification simulations, the molten metal temperature inside the mould was also measured for each laboratory weld. It was observed that the temperature of the molten metal after the pouring stage was not uniform throughout

[3] The term laboratory weld means that the experiment was performed under controlled conditions.

the weld cross-section. Therefore, for each laboratory weld, a polynomial curve was fitted to the measured temperatures inside the mould from the rail base to the rail-head. The resulting polynomial curve was defined as the corresponding laboratory weld's initial molten metal temperature profile. The experimentally obtained molten metal temperature profile was utilised as an initial condition in the simulations for the associated laboratory weld under investigation. Thus, the pouring stage was not exclusively simulated; instead, it was assumed that the mould was already filled with molten metal. Subsequently, the rail end inside the mould was assumed as a flat surface, i.e., the shape of the rail end inside the mould is not affected during the pouring stage. Moreover, they included the convection heat transfer; for this purpose, the effective thermal conductivity of liquid steel was enhanced by a factor of five. The authors validated their proposed model and approach against all the laboratory welds conducted in their study. The predicated FZ and HAZ have reasonably agreed with the experimental data. Furthermore, they performed a detailed sensitivity analysis to understand the influence of various parameters, such as preheating time, initial molten metal temperature, weld gap and ambient temperature, on the thermal conditions, i.e., FZ, HAZ, solidification time, etc. Their results show that the influence of the weld gap on the thermal conditions is significant. If the weld gap is unchanged, then the preheating time and the initial temperature of the molten metal are the potential parameters to influence the thermal conditions of a Thermit weld. The authors also performed a parametric analysis to predict the formation of known welding defects, such as shrinkage, micro-porosity cold-lap and centreline defects. They concluded that, although most defects can be eliminated or reduced by tweaking the welding parameters, the micro-porosity cannot be completely eliminated. Their results predicted that with an increase in preheating time and initial molten metal temperature, the formation of cold-lap and the centreline defect might be reduced; nonetheless, it is not as effective as increasing the weld gap. Finally, they proposed a weld gap of 38 mm with a slight increase in preheating time to obtain an optimum weld [15].

In 2011, Tuchkova [9] performed multiphase numerical simulations of the complete AT welding process (excluding AT-reaction and steel-slag separation in the crucible and preheating of the casting system) for the joining of rails. The simulations were carried out using commercial CFD code Flow3D®. The SkV procedure of the AT-Welding process, comprising an SkV-Standard mould assembly for the weld gap of 25 mm and a UIC60 rail profile, was investigated. In the SkV procedure, the complete casting system is preheated for 150 s using an oxy/propane torch, which is placed right above the rail head, followed by 30 s of tapping stage (i.e., the time required to insert the sand plug in the mould, positioning of the crucible, igniting the Thermit-mixture and the separation of the steel and slag inside the crucible).

However, the preheating and tapping stage was not directly simulated in the scope of her work. This was performed by an external institute[4] and the numerically obtained temperature field in the rail and casting mould after the preheating and tapping stage was used as initial temperature conditions in her multiphase simulations. For the mould filling simulations, turbulence model[5] was used, as the flow field inside the mould was turbulent due to the intricate shape of the mould. To validate the numerical model, a series of laboratory experiments were conducted. Holes were drilled in the rails at various positions to record the temporal evolution of the temperature utilising two kinds of thermocouples, viz., Type S (PtRh(10%)-Pt) and Type K (NiCr-Ni). The former were positioned close to the rail ends inside the mould to withstand the harsh temperature conditions during the mould filling, and the latter was positioned away from the rail ends. Moreover, a thermal camera was used to record the surface temperature of the molten metal during the mould-filling stage at the outlet of the crucible.[6] The laboratory welds were then prepared for metallographic analysis to obtain the maximum width of the fusion and heat-affected zone. In addition, the hardness and the microstructure formation along the longitudinal cross-section of the weld region (i.e., FZ and HAZ) were also investigated and discussed in her work. The numerically predicted temporal evolution of the temperature at prescribed locations has shown good agreement when compared with the experimental data. Although the maximum width of the FZ and HAZ have shown a similar trend/profile as observed in the experiments, a notable deviation was observed in the maximum width of the fusion zone in the head region. The numerically obtained maximum width of the fusion zone was underestimated significantly.[7] This was later attributed to the fact that free convection was not considered in the molten metal flow to minimise the computational resources required. [8]

Weiss et al. [16] studied the discharge process of the multiphase system from two types of crucibles, which are widely used in the railway industry for AT welding of rails, viz., multi- and single-use crucibles. The multi-use crucible is conical and

[4] The investigations carried out at the CHAIR OF TECHNICAL THERMODYNAMICS, TU BERGAKADEMIE FREIBERG. The simulations were carried out using the commercial code Star-CCM+. The temperature information was then imported in Flow3D®.

[5] The type of turbulence model used in her study was not mentioned.

[6] The thermal images for both single and multi-use crucibles were evaluated.

[7] A quantitative analysis comparing the numerical results with experiments for the maximum width of the fusion and heat-affected zones were not found in her work; instead, the results were discussed based on qualitative analysis

[8] This thesis was published in the German language. The summary and findings presented in this study were derived by translating research reported in the original work of the author using google translate.

made of waterglass-bonded aluminium oxide (wg-Al_2O_3) composite. In contrast, the single-use crucible (a.k.a Euro crucible) is cylindrical and is made of waterglass-bonded silicium oxide (wg-SiO_2) composite. The simulations were performed using Finite Volume Method (FVM) based open-source CFD library, OpenFOAM. The authors employed a multiphase, isothermal, incompressible, Volume of Fluid (VoF) based solver, `multiphaseInterFoam`. Additionally, they utilised the RANS $k-\varepsilon$ turbulence model as the flow was deemed highly turbulent at the outlet of the crucible. Subsequently, the authors conducted experiments to demonstrate the multiphase flow phenomena inside a single-use crucible using a water-oil-air system to validate the solver. For the numerical investigations, the volume of water and oil inside the crucible was similar to steel and slag. Their results show that the solver could predict the oil's penetration in water as observed in experiments. Furthermore, based on the findings of the water-oil-air system, the authors performed simulations to investigate the discharge process of the steel-slag-air system. They observed similar behaviour in the outflow. However, slag penetration in steel happens much later when compared to the water-oil-air system. In addition, the authors conducted experiments to determine the contact angle of the Thermit-steel on two types of composite materials used in the AT welding process [17]. The Euro crucible and casting mould are generally made of wg-SiO_2 composite, whereas the sand plug is made of wg-Al_2O_3-SiO_2 composite. The contact angle of Thermit-steel on wg-SiO2 and wg-Al_2O_3-SiO_2 composites was measured as $(131.4\pm3.5)^\circ$ and $(134.0\pm 5.6)^\circ$, respectively. The measurements were carried out in the temperature range of 1723.2 K–1978.9 K.

Weiss [18] extended his scope of research by numerically investigating the separation of Thermit-steel and slag inside the crucible during the AT reaction and the complete pouring stage (i.e., crucible discharge and mould filling) of the AT welding process. The simulations were carried out on an SkV-Elite mould assembly mounted around the gap (25 mm) between the rails to be joined. A UIC60 rail profile was used in the study. Subsequently, Weiss investigated the multiphase flow field inside the mould with and without the sand plug during the filling process. The numerical predictions showed that, on the one hand, if the sand plug is excluded during the welding process, not only high velocities in the steel phase are observed but also a high probability of slag entrapment in the gap within the rail region. On the other hand, if the sand plug is included, the probability of slag entrapment is reduced. However, large air pockets within the rail web regions are formed, which may also compromise the quality of the weld. Although the multiphase CFD simulations performed by the author provide detailed insights into the flow field during the pouring stage, the simulations were performed under the isothermal flow conditions, i.e., heat transfer and solid-liquid phase change were not considered.

Josefson and Bisschop [19] and Josefson et al. [20] studied the influence of additional forging on the FZ, HAZ and the residual stress field to reduce the width of HAZ and thus minimise the risk of weld defect formation. An additional forging methodology called ALFONS (ALigining, FOrging aNd Shearing) was implemented [21]. For this purpose, a sequential thermo-mechanical FE analysis was performed using the commercial software ABAQUS. The simulations were carried out on both 2D and 3D computational domains.[9] Thermal analysis: The simulation of different stages of the AT welding process, i.e., preheating, tapping, pouring, shearing and cooling, were conducted by employing a hybrid approach as adopted by Chen et al. [14], i.e., the experimental data[10] is utilised for not only validating the results but also serving as initial and boundary conditions. Unlike Chen et al. [15], they imposed two different heat-flux profiles as boundary conditions on the rail inside the mould, i.e., on the rail end and the rail running surface. Moreover, the authors did not consider the mould in their simulations but implemented it as a boundary condition. To simulate the pouring stage, the temperature of the molten filler metal inside the mould was linearly increased up to 2050 °C ($\approx$ 2323 K) for 11 s. Mechanical analysis: The ALFONS module was implemented in ABAQUS based on the experimental set-up and information provided by Goldschmidt Holding GmbH. In the ALFONS module, approximately 850 kg of the load is distributed evenly over 887 mm on the rails on either side of the mould. After 120 s of the pouring stage, this load is linearly forged into the molten metal up to 6 mm from either side of the mould. The additional forging is applied for a duration of 180 s. Their simulation results show that although the ALFONS methodology successfully reduced the width of the FZ and HAZ, the resulting reduction was insignificant, and only marginal differences were observed in the residual stress field when compared with the results of conventional AT welding process. Furthermore, the authors observed that the ALFONS methodology might lead to higher plastic strains and increase the risk of hot cracking in the welded region.

Manzke et al. [22] developed a Finite Difference Method (FDM) based reduced numerical model to predict the heat transfer during the TW process. The computational domain was greatly simplified by first assuming a bilateral symmetry of the welding system and then considering a two-dimensional horizontal cross-section at the rail web region, including the molten metal, mould, lutting sand and rail. The heat

[9] For the 3D simulations, only one-fourth of the computational domain was considered due to two symmetry planes, i.e., along the longitudinal plane and across the weld gap

[10] The experimental data was provided by Thermit welding and Goldschmidt Holding GmbH. The AT welding set-up consists of an SKV-Elite mould assembly mounted on a UIC60 rail profile with a weld gap of 24 mm, and the complete setup is preheated using an oxy/propane torch for a duration of 180 s followed by 55 s of tapping.

transfer and temperature distribution in the rail and mould during the pre-heating, tapping and pouring stages were provided through initial and boundary conditions. The solidification process was modelled using the enthalpy method [23]. Moreover, the authors also considered heat transfer through thermal radiation between rail and mould in their reduced numerical model. The simulations were based on the SkV procedure of the TW process. The temperature distribution in the rail for various sub-processes, such as pre-heating, tapping, pouring and cooling, predicted by their reduced model, were compared with those obtained using OpenFOAM and were found to be in good agreement.[11] Subsequently, the authors performed a sensitivity analysis using their developed model to investigate the influence of various parameters on the width of the FZ. However, the sensitivity analysis was performed on the SoV process of the TW process. Their results show that the temperature of the molten metal (i.e., liquid steel) and specific heat capacity of the rail steel have significant influence on the width of the FZ.

Later, Manzke [25] also performed three dimensional numerical simulations of the TW process. The simulations were carried out using Star-CCM+. The welding system investigated in his study was based on the Standard-SkV mould for UIC60 type rail profile. Taking the advantage of the bilateral symmetry only one-quarter of the domain was considered for the simulations. Three different heat transfer models were simulated. The results were compared with the available experimental results, i.e., maximum width of FZ and HAZ. In model 1, the heat transfer model was based on heat conduction only and the mould was implemented as boundary condition, whereas in model 2 (referred as model III in [25]) the natural convection[12] in the molten metal was also considered. Finally, model 3 (referred as model IV in [25]) was based on coupled conjugate heat transfer simulations i.e., convection in the molten metal and the thermal contact between the components such as, lutting sand-mould, mould-rail, mould-molten metal were considered in the simulations. The later also served as the validation case for the developed reduced numerical model in his work.[13]

Galay and Ilinykh [27] explored TW for joining and repairing railway tracks, examining the standard weld gap of 25 mm and considering a potential increase

[11] Manzke et al. [22] performed the numerical simulations using the solver `castFoam`, which was developed by Miehe [24] to investigate the horizontal twin-roll casting of the magnesium alloy.

[12] The first model to investigate the natural convection in the molten metal within the mould was mentioned in MOTEVAS Project [26].

[13] This thesis was published in the German language. The summary and findings presented in this study were derived by translating research reported in the original work of the author using google translate.

to 50 mm during rail repairs. For this purpose, the commercially available casting software, NovaFlow&Solid CV was used to conduct a comprehensive modelling study for predicting the impact on varying the weld gap on the maximum width of the FZ and HAZ. The simulations for the preheating stage were not exclusively conducted, instead, both the rails and the mould were initialised with a temperature of 850 °C ($\approx$ 1123 K). Furthermore, the assumption was made that pouring occurred instantaneously, with the temperature of the poured metal set at 2300 °C ($\approx$ 2573 K) Their findings revealed that an increase in the weld gap leads to an increase in the maximum width of the FZ by 30–60 %, thereby increasing the vulnerability of the FZ to rail fractures.

Liu et al. [28] developed a structural fatigue life prediction model of Thermit welded rail joints. This model considered factors such as weld-induced residual stress and stress relaxation due to external cyclic loads. They utilised a 3D FE model that included sequentially coupled thermo-mechanical simulations and cyclic plasticity constitutive simulation to analyse welding-induced residual stress and the redistribution of stresses during cyclic bend fatigue loading. The numerical model was developed in the commercially available software ABAQUS. Thermal simulations were performed and one of the heat-flux profiles reported by Chen et al. [13] was employed to predict the temperature distribution in the rail during the preheating stage. Moreover, the pouring stage was considered to be instantaneous with a pouring temperature of 2050 °C ($\approx$ 2323 K) and was modelled using element death and rebirth technique to predict the cooling stage. Their results showed the importance of considering residual stress and stress relaxation in predicting the fatigue life of the joint, and neglecting these factors could lead to an overestimation of the predicted life by at least 50 % in rail web to base location.

Lima [29] proposed 3D thermo-elastoplastic numerical model to simulate the AT welding process. The main objective of their work was to model all stages of the AT welding process for rails and identify key parameters affecting residual stress distribution. Similar to Liu et al. [28], they utilised the element birth and death technique to simulate the pouring stage. However, they went further by emulating the filling stage in the pouring region by activating elements initialised with pouring temperature in multiple layers from the bottom to the top of the mould over the span of the pouring time recorded in the experiments. Lima employed the commercially available software, Ansys APDL (Ansys Parametric Design Language) for the numerical investigations and validated the proposed model with the works of Josefson et al. [20]. Moreover, he extended his analysis to different rail types, namely, TR57[14]

[14] Manufactured in accordance with the American standard (AREMA), the TR57 has a mass of 57kg m^{-1}, while the TR68 has a mass of 67.50kg m^{-1}

and TR68[14] exploring the impact of various parameters, such as, type of rail steel alloy, preheating time and gap size on the distribution of the residual stress during the entire process. Their results show that, for TR68 rails, the larger weld gap resulted in the highest residual stress and a gap of 25 mm had no significant impact on residual stress with varying preheating times. For TR57 rails, changing the type of steel, particularly the micro-alloyed investigated in their study, intensified residual stress.[15]

2.3 Contribution of this PhD Thesis

The literature review conducted in this chapter is summarised in Table 2.1. It is evident that certain aspects, particularly the thermal pouring stage and solidification stage, are not fully explored, as the majority of the research reported in the literature focuses on the residual stress induced during the AT welding process and post-weld heat treatment. Moreover, the numerical models developed, proposed or employed are based solely on conduction-type heat transfer, and the pouring stage is often skipped, given its complexity and computational expense. However, understanding the influence of the pouring and cooling (solidification) stages during the AT welding process is of utmost importance in obtaining an efficient weld, as poor execution of these stages may result in shrinkage defects, thus, adversely affecting weld quality and consequently influencing the residual stresses. With these gaps acknowledged, the current work focuses on comprehensively exploring the influence of various parameters on the thermal pouring and solidification stages. Specifically, the complexity of the thermal pouring stage, which has not been undertaken in previous literature, is emphasised.

The primary contribution of this PhD thesis lies in the development of a numerical model predicting the thermal flow field during the pouring and cooling stages of the AT welding process. This model is developed and implemented within the FVM based open-source CFD software, OpenFOAM, utilising the existing multiphase solver, `interFoam`, which is based on the VoF method. The sequence of extensions to the existing solver is as follows:

[15] This thesis was published in the Portuguese language. The summary and findings presented in this study were derived by translating research reported in the original work of the author using google translate.

Table 2.1 Overview of the literature review discussed in this chapter. The description of the abbreviations are as follows: C—Conduction, CC—Conduction and Convection, I—Isothermal, HFP—Heat Flux Profile, Sim—Simulation

Authors	Year	Domain	Heat transfer Model	Numerical Method	Simulation tool	Pouring stage	Pouring temperature	Solidification/ Melting	Preheating
Jha [8]	1989	3D	C	FEM	–	No	2948 K–2848 K	Yes	HFP
Chen et al. [13]	2004	3D	C	FEM	FIDAP	No	2073 K–2473 K	Yes	HFP
Chen et al. [14, 15]	2006								
Tuchkova [9]	2011	3D	C		Flow3D	Yes	2473 K	Yes	Sim
Manzke et al. [22]	2017	2D	C	FDM	C/C++	No	2473 K–2773 K	Yes	Sim
Weiss et al. [16]	2018	3D	I	FVM	OpenFOAM	Yes	2473 K	No	No
Weiss [18]	2019								
Josefson et al. [20]	2020	3D	C	FEM	ABAQUS	No	2323 K	No	HFP
Manzke [25]	2021	1D,3D	CC	FVM	Star-CCM+	No	2473 K	Yes	Sim
Galay and Ilinykh [27]	2021	3D	C	–	NovaFlow	No	2573 K	No	I
Liu et al. [28]	2021	3D	C	FEM	ABAQUS	No	2323 K	No	HFP
Lima [29]	2023	3D	C	FEM	ANSYS	–	2323 K	Yes	HFP
Present Work	**2024**	**3D**	**CC**	**FVM**	**OpenFOAM**	**Yes**	2073 K–2673 K	**Yes**	**I, HFP**

- The initial step involves the incorporation of the energy conservation equation to predict the heat transfer within the system.
- Subsequently, the enthalpy-porosity model is implemented to predict the solid-liquid phase change phenomenon within the system.
- Finally, recognising the turbulent nature of the pouring stage [9, 18], the $k - \varepsilon$ turbulence model is employed in this study. Additionally, the turbulence model is adopted to account for the solid-liquid phase change phenomenon, enabling faster computations.

The developed numerical model is rigorously validated through well-documented benchmark cases reported in the literature. Additionally, in-situ experiments are conducted to capture the temporal and spatial evolution of solid-liquid interface fronts during the melting of wax within an enclosure in the presence of gas phase. This experimental data is then compared with the numerical predictions, leading to a comprehensive discussion on deviations and limitations of the developed numerical model.

As extremely high temperatures are involved during the AT welding process, and due to the opaqueness of the welding set-up, novel experiments were conducted. These experiments aim to obtain the shape of the melted rails within the mould post-pouring stage. The experimentally obtained melted rails are digitalised using photogrammetry to capture intricate shape of the melted regions, providing a valuable resource for the verification of numerical predictions pertaining to the thermal pouring stage. Furthermore, additional experiments are performed to obtain the temporal and spatial evolution of the temperature field within the rails, followed by the final weld profiles.

The validated numerical model is employed to predict the thermal flow field and the melting of rails within the mould during the pouring stage. The predicted shape of the melted rails are verified by comparing them with the digitalised version. This thesis also delves into the exploration of the influence of various pouring temperatures on melted rail profiles and the investigation of the impact of various molten metal temperatures, thermophysical properties, and the inclusion of a preheating stage on final weld profiles.

3 Numerical Modelling

This chapter provides a comprehensive overview of the numerical model developed for predicting the flow field, heat transfer and the solid-liquid phase change during the mould-filling and cooling stage of the AT welding process. The developed numerical model shall be implemented in the open-source CFD toolkit OpenFOAM utilising the `interFoam` solver as its foundation. This solver employs the Volume of Fluid (VoF) method to capture the interface between the fluids. The solver is further extended by introducing an additional Partial Differential Equation (PDE) to incorporate heat transfer within the multiphase (solid, liquid and gas) system. Furthermore, the enthalpy-porosity approach is adopted for predicting the temporal and spatial evolution of the solid-liquid interface front. The multiphase turbulence model is discussed and slightly modified to account for the damping of the turbulence parameters within the mushy and solid regions. Additionally, the chapter provides insights into the Finite Volume Method (FVM) within OpenFOAM and concludes by demonstrating the implementation/representation of a PDE in the OpenFOAM framework.

3.1 Governing Equations

The equations of conservation for mass (a.k.a. continuity equation), momentum (a.k.a. Navier-Stokes equations) and energy for a fluid passing through an infinitesimal fixed control volume (CV) form the fundamental set of equations. These equations describe the dynamics of heat and fluid flow, where the fluid is regarded as a continuum. The unsteady governing equations in conservative form read as follows:

R. G. Kewalramani, *Computational Thermo-Fluid Dynamics of Aluminothermic Welding Process*, https://doi.org/10.1007/978-3-658-47535-2_3

Conservation of Mass:

$$\frac{\partial \varrho}{\partial t} + \nabla \cdot (\varrho \boldsymbol{u}) = 0, \tag{3.1}$$

Conservation of Momentum:

$$\frac{\partial (\varrho \boldsymbol{u})}{\partial t} + \nabla \cdot (\varrho \boldsymbol{u}\boldsymbol{u}) = -\nabla p + \nabla \cdot \boldsymbol{\tau}_{ij} + \boldsymbol{S}_b + \boldsymbol{S}_u, \tag{3.2}$$

Conservation of Energy:

$$\frac{\partial (\varrho h)}{\partial t} + \nabla \cdot (\varrho \boldsymbol{u} h) = -\nabla \cdot \dot{\boldsymbol{q}} + \boldsymbol{S}_h. \tag{3.3}$$

Here ϱ is the density, t is the time, $\boldsymbol{u}$ is the velocity vector, $\boldsymbol{\tau}_{ij}$ is the viscous stress tensor, $\boldsymbol{S}_b$ represents the buoyancy source term, $\boldsymbol{S}_u$ represents sources and sinks in the momentum equation, h is the specific enthalpy, $\dot{\boldsymbol{q}}$ is the heat flux vector, $\boldsymbol{S}_h$ represents sources and sinks in energy equation.

The following assumptions are made to simplify the system of equations to predict the flow field and heat transfer during the AT welding process:

1. The fluid is treated as incompressible
   Equation (3.1) represents the general form of mass conservation equation. The rate of change of density is accounted for by the first term, while the net flow of mass (or fluid volume) out of the surface of the CV is described by the second term. Equation (3.1) can be re-written[1], such that:

   $$\frac{\partial \varrho}{\partial t} + \boldsymbol{u} \cdot \nabla \varrho + \varrho \nabla \cdot \boldsymbol{u} = 0. \tag{3.4}$$

   For an incompressible fluid, the density of the fluid is assumed to be constant in time and space. As a result, the temporal and spatial derivatives of density can be eliminated from the equation (3.4) and the mass conservation equation is reduced to:

   $$\nabla \cdot \boldsymbol{u} = 0. \tag{3.5}$$

2. The fluid is considered Newtonian:
   For Newtonian fluids the viscous stress tensor depends on the divergence of the velocity field and strain rate and takes the form [30]:

---

[1] Using vector identity: $\nabla \cdot (s\mathbf{v}) = \mathbf{v} \cdot \nabla s + s\nabla \cdot \mathbf{v}$, where s is a scalar and $\mathbf{v}$ is a vector.

$$\boldsymbol{\tau}_{ij} = 2\mu \boldsymbol{S}_{ij} - \left(\frac{2}{3}\mu \nabla \cdot \boldsymbol{u}\right)\delta_{ij}, \tag{3.6}$$

where μ is the dynamic viscosity, δ_{ij} is the Kronecker delta function (δ_{ij} is 1 for $i = j$ and 0 for $i \neq j$) and $\boldsymbol{S}_{ij}$ is the strain rate tensor (i.e., deformation tensor):

$$\boldsymbol{S}_{ij} = \frac{1}{2}(\nabla \boldsymbol{u} + (\nabla \boldsymbol{u})^T), \tag{3.7}$$

Combining equations (3.6) and (3.7) yields the viscous stress tensor equation for Newtonian fluids:

$$\boldsymbol{\tau}_{ij} = \mu\big(\nabla \boldsymbol{u} + (\nabla \boldsymbol{u})^T\big) - \left(\frac{2}{3}\mu \nabla \cdot \boldsymbol{u}\right)\delta_{ij}. \tag{3.8}$$

Finally for an incompressible fluid, the stress tensor equation (3.8), can be reduced to:

$$\boldsymbol{\tau}_{ij} = \mu\big(\nabla \boldsymbol{u} + (\nabla \boldsymbol{u})^{\mathrm{T}}\big). \tag{3.9}$$

3. Material properties are assumed to be constant:
   There exists a lack of information in the literature regarding the material properties of Thermit-steel above its liquidus temperature. Hence, for simplifying the computations, the material properties of each phase and fluid involved in the system, such as density (ϱ), dynamic viscosity (μ), thermal conductivity (κ) and specific heat capacity (c_p) are considered constant values and independent of temperature[2]. However, density variations caused by changes in temperature is considered only in the buoyancy source term through Boussinesq approximation and shall be discussed in Sect. 3.3.
4. Fourier law of heat conduction:
   The heat flux vector ($\dot{\boldsymbol{q}}$) is expressed in terms of temperature gradient (∇T) by invoking the Fourier law of heat conduction and reads:

   $$\dot{\boldsymbol{q}} = -\kappa \nabla T, \tag{3.10}$$

   were κ is the thermal conductivity.
5. For the sake of simplicity, radiative heat transfer is not taken into account.

[2] It is important to note that this approach may not provide the most accurate results. Therefore, further investigations to obtain/measure the required material properties may be necessary to achieve more precise results.

3.2 Volume of Fluid Method

The prediction of the dynamics of the multi-fluid[3] system involved during the AT welding process is accomplished by employing the Volume of fluid (VoF) method. In the VoF method [31], an indicator or colour function is employed to determine the volume fraction of a fluid within each control volume (CV) of the computational domain and is defined as:

$$\alpha_i = \frac{V_i}{V_{\mathrm{CV}}}, \tag{3.11}$$

where α is the volume fraction of the fluid i, the volume of the fluid V_i in a CV and V_{CV} the total volume of the CV. The volume fraction $\alpha_i \in [0, 1]$ and is defined as:

$$\alpha_i = \begin{cases} 1 & \text{cell contains fluid 1,} \\ 0 < \alpha_i < 1 & \text{cell contains interface,} \\ 0 & \text{cell contains fluid 2.} \end{cases} \tag{3.12}$$

Figure 3.1 illustrates the volume fraction field of the VoF method within the computational domain during the mould filling stage of the AT welding of rails. It is a scalar variable and is associated with each fluid in order to identify different fluids in the computational domain (Figure 3.1). Therefore, the conservation equation for transport of a scalar quantity α_i is of the form [31]:

$$\frac{\partial \alpha_i}{\partial t} + \nabla \cdot (\alpha_i \boldsymbol{u}) = 0, \tag{3.13}$$

where $\boldsymbol{u} = \boldsymbol{u}_1 = \boldsymbol{u}_2$. Herein it is sufficient, if the volume fraction of one fluid is known. The volume fraction of the other fluid is computed by:

$$\begin{aligned} \sum_{i=1}^{2} \alpha_i &= 1, \\ \alpha_1 + \alpha_2 &= 1, \\ \alpha_2 &= 1 - \alpha_1. \end{aligned} \tag{3.14}$$

[3] In discussions related to the VoF method, "multi-fluid" or "two-fluid" are used instead of "multiphase" or "two-phase" to ensure clarity. Subsequently, the solid-liquid transformations within the fluid are referred to as "phases", which shall be discussed in Sect. 3.3.

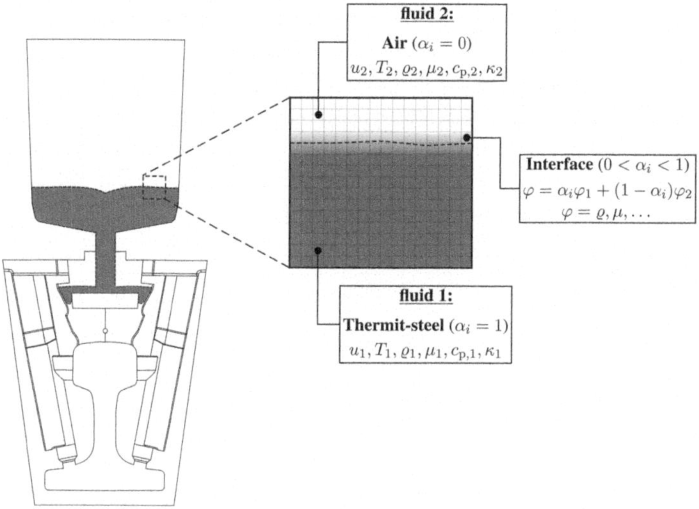

Figure 3.1 A schematic representation of the volume fraction field in the VoF method, depicting the interface between liquid Thermit-steel (grey) and air (white)

The utilisation of the VoF method makes it possible to use only one set of governing equations for the entire flow field in the computational domain resulting in a single-field formulation. Additionally, the local material property ϕ of the fluid, which is discontinuous across the interface, is described by the weighted average of the volume fraction of each fluid in the CV and is given by:

$$\phi = \alpha_1\phi_1 + \alpha_2\phi_2. \tag{3.15}$$

This method is computationally inexpensive [31], as only one storage variable, i.e., α for each CV is required in order to define the interface.

Although the VoF method is the most popular choice amongst the researchers to solve a multi-fluid system, the accuracy of the method is not high and a diffusive interface is often the result, wherein the interface is smeared over a few cells. This is directly associated to the discretisation of the convection term. In order to address this issue, alternative methods have been developed. These can be categorised into direct (such as the Compressive Interface Capturing Scheme for Arbitrary Meshes (CICSAM) [32, 33] and High Resolution Interface Capturing (HRIC) Scheme [34]), reconstruction (such as the Piecewise Linear Interface Calculation (PLIC) [35] and Simple Line Interface Calculation (SLIC) [36]) methods [37]. A few hybrid (such as the Moment of Fluid (MOF) [38], Coupled Level Set and Volume of Fluid (CLSVOF) [39] and Interface Reconstruction Coupled Level Set and Volume of Fluid (IRCLS-VOF) [40]) methods have also been developed. In this study, the so-called Weller scheme, which incorporates an additional convective term in the equation (3.13) in order to minimise the numerical diffusion at the interface, is employed, as discussed by Rusche [41].

Reformulated VoF Method

The VoF equation (3.13) for each fluid in a two-fluid system can be expressed as:

$$\frac{\partial \alpha_1}{\partial t} + \nabla \cdot (\alpha_1 \boldsymbol{u}_1) = 0, \tag{3.16}$$

$$\frac{\partial \alpha_2}{\partial t} + \nabla \cdot (\alpha_2 \boldsymbol{u}_2) = 0. \tag{3.17}$$

In contrast to the original VoF method, the velocity of the effective fluid is now defined as the weighted average of phase velocities and reads:

$$\boldsymbol{u} = \alpha_1 \boldsymbol{u}_1 + \underbrace{(1 - \alpha_1)}_{\alpha_2} \boldsymbol{u}_2 \tag{3.18}$$

Re-arranging equation (3.16), we get:

$$\frac{\partial \alpha_1}{\partial t} + \nabla \cdot (\alpha_1 \boldsymbol{u}_1) + \big[\underbrace{\nabla \cdot (\alpha_1 \boldsymbol{u}) - \nabla \cdot (\alpha_1 \boldsymbol{u})}_{=0}\big] = 0, \tag{3.19}$$

$$\frac{\partial \alpha_1}{\partial t} + \nabla \cdot (\alpha_1 \boldsymbol{u}) + \big[\nabla \cdot (\alpha_1 \boldsymbol{u}_1 - \alpha_1 \boldsymbol{u})\big] = 0. \tag{3.20}$$

Replacing $\boldsymbol{u}$ in the third term of equation (3.20) with equation (3.18) and re-arranging terms:

$$\frac{\partial \alpha_1}{\partial t} + \nabla \cdot (\alpha_1 \boldsymbol{u}) + \big[\nabla \cdot (\alpha_1 \boldsymbol{u}_1 - \alpha_1(\alpha_1 \boldsymbol{u}_1 + (1 - \alpha_1)\boldsymbol{u}_2))\big] = 0, \tag{3.21}$$

$$\frac{\partial \alpha_1}{\partial t} + \nabla \cdot (\alpha_1 \boldsymbol{u}) + \big[\nabla \cdot (\alpha_1 \boldsymbol{u}_1 - {\alpha_1}^2 \boldsymbol{u}_1 - \alpha_1(1 - \alpha_1)\boldsymbol{u}_2)\big] = 0, \tag{3.22}$$

$$\frac{\partial \alpha_1}{\partial t} + \nabla \cdot (\alpha_1 \boldsymbol{u}) + \big[\nabla \cdot (\alpha_1(1 - \alpha_1)\boldsymbol{u}_1 - \alpha_1(1 - \alpha_1)\boldsymbol{u}_2)\big] = 0, \tag{3.23}$$

$$\frac{\partial \alpha_1}{\partial t} + \nabla \cdot (\alpha_1 \boldsymbol{u}) + \left[\nabla \cdot (\alpha_1 (1-\alpha_1) \underbrace{(\boldsymbol{u}_1 - \boldsymbol{u}_2)}_{\boldsymbol{u}_\text{r}})\right] = 0 \tag{3.24}$$

where $\boldsymbol{u}_\text{r} = \boldsymbol{u}_1 - \boldsymbol{u}_2$, is the relative velocity between the two fluids. The third term in equation (3.24) is the artificial compression term used to compress the interface between the two fluids and has no physical significance. The term is active only in the vicinity of the interface due to its dependence on $\alpha_1(1-\alpha_1)$, where $\alpha_1 \in (0, 1)$ and disappears in individual fluids.

Surface tension force

Surface tension force exists at the interface for any pair of fluids for e.g air-water, water-oil, steel-slag, steel-air, etc. Thus, creating a pressure jump across the interface and the magnitude of this pressure jump (Δp) is expressed by Young-Laplace equation, which reads [32]:

$$\Delta p = p_1 - p_2 = \sigma k, \tag{3.25}$$

where σ is the surface tension coefficient and is defined as the amount of work required to create a unit area of free surface, its value is positive for immiscible fluids and its magnitude is determined by the nature of the fluids. The mean curvature k of the interface, given by:

$$k = \left(\frac{1}{R_1} + \frac{1}{R_2}\right), \tag{3.26}$$

where R_1 and R_2 are the principal curvature radii of the surface.

In the VoF method, the interface is not tracked explicitly, and subsequently, the exact shape and location are unknown [41]. Additionally, the pressure jump exists only within the transition region and is discontinues. Therefore, the implementation and evaluation of the surface tension force in the momentum equation presents a challenge. However, in 1992, Brackbill et al. [42] solved this problem by proposing a Continuum Surface Force (CSF) model, which represents the surface tension effects as a continuous volumetric force that acts within the transitional region of the interface. It reads:

$$(S_{\boldsymbol{u}} =)\ S_\sigma = \sigma k \nabla \alpha_1, \tag{3.27}$$

where the gradient $\nabla \alpha$ is a continuous function which is active (i.e., non zero) only in the transitional region and k is the interfacial curvature and is expressed as:

$$k = -\nabla \cdot \boldsymbol{n}, \tag{3.28}$$

where $\boldsymbol{n}$ is the interface unit normal vector and is defined as:

$$\boldsymbol{n} = \frac{\nabla \alpha_1}{|\nabla \alpha_1|}. \tag{3.29}$$

3.3 Solid-liquid Phase Change Model

In order to predict the temporal and spatial evolution of the solid-liquid interface front during the AT welding process, the enthalpy-porosity approach [43, 44] is adopted. By applying the assumptions outlined in Sect. 3.1 the energy equation (3.3) reduces to:

$$\frac{\partial(\varrho h)}{\partial t} + \nabla \cdot (\varrho \boldsymbol{u} h) = \nabla \cdot (\kappa \nabla T) + \boldsymbol{S}_h. \tag{3.30}$$

Following Sect. 3.2, the specific enthalpy in equation (3.30) for a two-fluid system can be expressed using equation (3.15) as:

$$h = \alpha_1 h_1 + \alpha_2 h_2, \tag{3.31}$$

where h_1 and h_2 represent the specific enthalpy of fluid 1 and fluid 2, respectively. The solid-liquid phase change is modelled within fluid 1, the specific enthalpy h_1 can be calculated using the weighted average of the melt fraction γ of each phase [23]:

$$h_1 = \gamma h_{1,\mathrm{l}} + (1 - \gamma) h_{1,\mathrm{s}}, \tag{3.32}$$

where subscripts s and l denote solid and liquid phase, respectively. The melt fraction $\gamma \in [0, 1]$ is defined as a linear function of temperature between solidus and liquidus temperature, T_s and T_l, respectively, and is expressed as:

$$\gamma(T) = \begin{cases} 0 & T < T_\mathrm{s} & \text{solid,} \\ \frac{T - T_\mathrm{s}}{T_\mathrm{l} - T_\mathrm{s}} & T_\mathrm{s} \leq T \leq T_\mathrm{l} & \text{mushy zone,} \\ 1 & T > T_\mathrm{l} & \text{liquid.} \end{cases} \tag{3.33}$$

Combining equations (3.31) and (3.32), the total specific enthalpy h of the multi-fluid system including solid-liquid phase change can be defined as:

$$h = \alpha_1 \left[\gamma h_{1,\mathrm{l}} + (1 - \gamma) h_{1,\mathrm{s}} \right] + \alpha_2 h_2. \tag{3.34}$$

The specific enthalpy for each fluid is given by:

$$h_1 = \gamma\Bigg[\underbrace{\int_{T_{\mathrm{ref}}}^{T} c_{p,1,1}\,\mathrm{d}T + \mathcal{L}}_{h_{1,1}}\Bigg] + (1-\gamma)\Bigg[\underbrace{\int_{T_{\mathrm{ref}}}^{T} c_{p,1,\mathrm{s}}\,\mathrm{d}T}_{h_{1,\mathrm{s}}}\Bigg], \tag{3.35}$$

$$h_2 = \int_{T_{\mathrm{ref}}}^{T} c_{p,2}\,\mathrm{d}T, \tag{3.36}$$

where c_p is the specific heat, $\mathcal{L}$ the latent heat of fusion and T_{ref} an arbitrary reference temperature. Assuming that the material properties are constant for each phase and fluid equations (3.35) and (3.36) can be integrated as follows:

$$h_1 = \gamma\Big[c_{p,1,1}(T - T_{\mathrm{ref}}) + \mathcal{L}\Big] + (1-\gamma)\Big[c_{p,1,\mathrm{s}}(T - T_{\mathrm{ref}})\Big], \tag{3.37}$$

$$h_2 = c_{p,2}(T - T_{\mathrm{ref}}). \tag{3.38}$$

Substituting equations (3.37) and (3.38) in equation (3.34) yields the general form of specific enthalpy for the multiphase system as a function of temperature:

$$h(T) = c_p(T - T_{\mathrm{ref}}) + \alpha_1\gamma\mathcal{L}. \tag{3.39}$$

Combining equations (3.39) and (3.3), the energy conservation equation in terms of temperature is obtained:

$$\begin{aligned}\frac{\partial(\varrho c_p T)}{\partial t} + \nabla\cdot(\varrho \boldsymbol{u} c_p T) = \nabla\cdot(\kappa\nabla T) \underbrace{- \alpha_1\mathcal{L}\Bigg[\frac{\partial\varrho\gamma}{\partial t} + \nabla\cdot(\varrho\boldsymbol{u}\gamma)\Bigg]}_{S_h} \\ + T_{\mathrm{ref}}\Bigg[\frac{\partial\varrho c_p}{\partial t} + \nabla\cdot(\varrho\boldsymbol{u}c_p)\Bigg].\end{aligned} \tag{3.40}$$

The material properties of the complete multiphase system are expressed in terms of volume fraction α_1 and melt fraction γ, and is defined as:

$$\phi = \alpha_1\big[\underbrace{\gamma\phi_1 + (1-\gamma)\phi_{\mathrm{s}}}_{\phi_1}\big] + \alpha_2\phi_2, \tag{3.41}$$

where ϕ is either ϱ, c_p, κ, etc.

Boussinesq approximation (Natural convection)

The Boussinesq approximation [45] provides a simplified approach for solving problems associated to natural convection that takes into account the buoyancy effects caused by temperature gradients within the fluid. This approach assumes that the density change in the fluid due to temperature variations is small enough to be ignored in the inertia terms of the Navier-Stokes equations, but significant enough to be included in the terms accompanied by gravitational constant. By considering this simplification, the Boussinesq approximation eliminates the need to solve the complete compressible formulation of the Navier-Stokes equations. Thus, with the Boussinesq approximation, the buoyancy term reads [30, 46]:

$$S_b = \varrho \mathbf{g} = \varrho_0 \mathbf{g}[1 - \beta(T - T_0)], \tag{3.42}$$

where $\mathbf{g} = (g_x, g_y, g_z)$ is the gravity vector, ϱ_0 is the fluid density at temperature T_0 and β is the thermal expansion coefficient.

Velocity field in solid phase

In order to modify the velocity field in the mushy region and in the solid phase, a D'Arcy type source term is introduced into the momentum conservation equation (3.2). This source term ensures that the velocity field in the solid phase within the computational domain is enforced to zero, while treating the mushy zone as a porous medium. To achieve this, the Carman-Kozeny equation [44, 47] for flows in porous media is adopted. The D'Arcy type source term reads:

$$(S_{\boldsymbol{u}} \mathrel{+}=)\; S_{\mathrm{D}} = \mathcal{A}(\gamma)\boldsymbol{u}. \tag{3.43}$$

Herein, $\mathcal{A}$ is the porosity function, which is determined by melt fraction (γ) and is computed by the following expression:

$$\mathcal{A}(\gamma) = -\alpha_1 \left[C \frac{(1-\gamma)^2}{\gamma^3 + \epsilon} \right], \tag{3.44}$$

where C is the large constant that describes the morphology of the mushy zone, $\epsilon = 1 \times 10^{-3}$ is the small computational constant to prevent a "divide by zero" error when melt fraction is zero and α_1 is introduced to restrict the term to fluid 1 only. In equation (3.44), when $\gamma = 0$ the term dominates in the CVs containing solid region, enforcing the velocity in those CVs to zero. Conversely, when $\gamma = 1$, the term vanishes in the CVs, allowing for free motion in the liquid region.

Therefore, the final form of the momentum equation for predicting the flow field within the mould during the mould-filling stage and solidification process is obtained by incorporating the previously discussed source terms, such as surface tension (equation (3.27)), natural convection (equation (3.42)), Darcy term (equations (3.43) and (3.44)) into the general form of the momentum equation (3.2) for Newtonian fluid and reads:

$$\frac{\partial(\varrho \boldsymbol{u})}{\partial t} + \nabla \cdot (\varrho \boldsymbol{u}\boldsymbol{u}) = -\nabla p + \nabla \cdot \mu\big(\nabla \boldsymbol{u} + (\nabla \boldsymbol{u})^T\big) + \underbrace{\varrho_0 \boldsymbol{g}[1 - \beta(T - T_0)]}_{S_b}$$
$$\underbrace{+ \sigma k \nabla \alpha_1 - \alpha_1 \left[C \frac{(1-\gamma)^2}{\gamma^3 + \epsilon}\right] \boldsymbol{u}}_{S_u} \quad (3.45)$$

3.4 Multiphase Turbulence Modelling

The equations governing multi-fluid, multiphase heat transfer and solid-liquid phase change as explored in Sects. 3.1–3.3, are applicable under conditions of laminar flow. Nevertheless, in the industrial scenarios like the AT welding process, the multiphase flow system within the crucible and mould typically experiences turbulent flow [9, 18]. Thus, in the subsequent step, the Reynolds-Average Navier-Stokes (RANS) equations are formulated and the $k - \varepsilon$ turbulence model is discussed with the aim of addressing the closure problem linked with turbulent flow simulations.

Reynolds-averaged conservation equations

The generic variable φ (i.e., $\boldsymbol{u}$, p, T, etc.) is calculated as the sum of its time-averaged value $\overline{\varphi}(\boldsymbol{x}, t)$ and its fluctuating value $\varphi\prime(\boldsymbol{x}, t)$ [46]:

$$\varphi(\boldsymbol{x}, t) = \overline{\varphi}(\boldsymbol{x}, t) + \varphi\prime(\boldsymbol{x}, t), \quad (3.46)$$

where $\boldsymbol{x} = (x, y, z)$ is the position vector.

On substituting equation (3.46) in continuity equation (3.5), VoF equation (3.24), momentum equation (3.45) and energy equation (3.40), the RANS equations are obtained:

$$\nabla \cdot (\overline{\boldsymbol{u}}) = 0, \quad (3.47)$$

$$\frac{\partial \alpha_1}{\partial t} + \nabla \cdot (\alpha_1 \overline{\boldsymbol{u}}) + \nabla \cdot (\alpha_1 (1 - \alpha_1) \overline{\boldsymbol{u}}_{\mathrm{r}}) = 0, \tag{3.48}$$

$$\frac{\partial (\varrho \overline{\boldsymbol{u}})}{\partial t} + \nabla \cdot (\varrho \overline{\boldsymbol{u}\boldsymbol{u}}) = -\nabla \overline{p} + \nabla \cdot \left[\mu\left(\nabla \overline{\boldsymbol{u}} + (\nabla \overline{\boldsymbol{u}})^T - \underbrace{\varrho \overline{\boldsymbol{u}'\boldsymbol{u}'}}_{\tau_{ij}^R}\right)\right] + \overline{\boldsymbol{S}}_b + \overline{\boldsymbol{S}_{\boldsymbol{u}}}, \tag{3.49}$$

$$\frac{\partial (\varrho c_p \overline{T})}{\partial t} + \nabla \cdot (\varrho \overline{\boldsymbol{u}} c_p \overline{T}) = \nabla \cdot (\kappa \nabla \overline{T} - \underbrace{\varrho c_p \overline{T'\boldsymbol{u}'}}_{\dot{\boldsymbol{q}}^R}) - \overline{S}_h. \tag{3.50}$$

However, the above Reynolds-averaged equations due to the turbulent motion of fluid flow, introduce additional unknown terms, $-\varrho \overline{\boldsymbol{u}'\boldsymbol{u}'}$ and $-\varrho c_p \overline{T'\boldsymbol{u}'}$. These terms represent the averaged products of fluctuating quantities due to the non-linear terms, representing the Reynolds stress tensor τ_{ij}^R and turbulent heat fluxes $\dot{\boldsymbol{q}}^R$, respectively.

The Reynolds stress tensor is a symmetric second order tensor and introduces six new unknown variables (comprising three normal and three shear stresses) into the momentum equation (3.49). Subsequently, three additional unknowns are introduced into the energy equation (3.50) by the turbulent heat fluxes. Consequently, this leads to a situation where the number of unknowns surpasses the number of equations, giving rise to what is commonly referred to as the turbulence closure problem.

The Reynolds stress tensor is modelled as a linear eddy viscosity model based on Boussinesq hypothesis, which assumes the Reynolds stress to be a linear function of the velocity gradients. Therefore, the Reynolds stress tensor for incompressible flows is given by [48]:

$$\tau_{ij}^R = -\varrho \overline{\boldsymbol{u}'\boldsymbol{u}'} = \mu_{\mathrm{t}}\left[\nabla \boldsymbol{u} + (\nabla \boldsymbol{u})^{\mathrm{T}}\right] - \frac{2}{3}\varrho k I, \tag{3.51}$$

where μ_t is the turbulent viscosity (or eddy viscosity) and depends on the local flow conditions, and k is the turbulent kinetic energy:

$$k = \frac{1}{2}\overline{\boldsymbol{u}'\boldsymbol{u}'} = \frac{1}{2}(\overline{{u_x}'^2} + \overline{{u_y}'^2} + \overline{{u_z}'^2}). \tag{3.52}$$

The turbulent heat fluxes are modelled akin to the momentum equation and are computed in terms of Fourier's law:

$$\begin{aligned}\dot{q}^R &= -\varrho c_p \overline{T\prime \boldsymbol{u}\prime} \\ &= \frac{\mu_\mathrm{t}}{Pr_\mathrm{t}} c_p \nabla T \\ &= \kappa_\mathrm{t} \nabla T,\end{aligned} \tag{3.53}$$

where κ_t is the turbulent thermal conductivity and is given by:

$$\kappa_t = \frac{c_p \mu_\mathrm{t}}{Pr_\mathrm{t}}, \tag{3.54}$$

where Pr_t is the turbulent Prandtl number, which also depends on the local flow conditions instead of the properties of the material. The final form of the momentum and energy equations can now be expressed as:

$$\frac{\partial(\varrho\overline{\boldsymbol{u}})}{\partial t} + \nabla\cdot(\varrho\overline{\boldsymbol{u}\boldsymbol{u}}) = -\nabla\overline{p} + \nabla\cdot\Big[\underbrace{(\mu+\mu_\mathrm{t})}_{\mu_\mathrm{eff}}\big(\nabla\overline{\boldsymbol{u}} + (\nabla\overline{\boldsymbol{u}})^\mathrm{T}\big)\Big] + \overline{\boldsymbol{S}}_b + \overline{\boldsymbol{S}}_{\boldsymbol{u}}, \tag{3.55}$$

$$\frac{\partial(\varrho c_p \overline{T})}{\partial t} + \nabla\cdot(\varrho\overline{\boldsymbol{u}} c_p \overline{T}) = \nabla\cdot[\underbrace{(\kappa+\kappa_\mathrm{t})}_{\kappa_\mathrm{eff}}\nabla\overline{T}] - \overline{S}_h. \tag{3.56}$$

Turbulence model

In this study, the standard $k-\varepsilon$ model [49] with additional source terms to account for solid-liquid phase change is adopted to describe the turbulence of the flow (adapted from [50, 51]). Herein, the mean velocity field [equation (3.55)] is solved along with transport equations for turbulent kinetic energy (k) and its dissipation rate (ε). The modified $k-\varepsilon$ model equations are :

$$\frac{\partial\varrho k}{\partial t} + \nabla\cdot(\varrho\overline{\boldsymbol{u}}k) = \nabla\cdot(\mathcal{D}_k\nabla k) + \mathcal{P}_k - \varrho\varepsilon + \boldsymbol{S}_{k,\mathrm{D}} \tag{3.57}$$

$$\frac{\partial\varrho\varepsilon}{\partial t} + \nabla\cdot(\varrho\overline{\boldsymbol{u}}\varepsilon) = \underbrace{\nabla\cdot(\mathcal{D}_\varepsilon\nabla\varepsilon)}_{I} + \underbrace{\mathcal{P}_\varepsilon}_{II} - \underbrace{C_{2\varepsilon}\varrho\frac{\varepsilon^2}{k}}_{III} + \underbrace{\boldsymbol{S}_{\varepsilon,\mathrm{D}}}_{IV} \tag{3.58}$$

The first term (I) on the right hand side of the equations (3.57) and (3.58) represents the diffusion term and the coefficients $\mathcal{D}_k$ and $\mathcal{D}_\varepsilon$ are expressed as:

$$\mathcal{D}_k = \mu + \frac{\mu_t}{\sigma_k}, \tag{3.59}$$

$$\mathcal{D}_\varepsilon = \mu + \frac{\mu_t}{\sigma_\varepsilon}, \tag{3.60}$$

where μ_t is defined as [49]:

$$\mu_t = C_\mu \varrho \frac{k^2}{\varepsilon}. \tag{3.61}$$

The second term $(II)$is the production term due to mean velocity gradients and the coefficients $\mathcal{P}_k$ and $\mathcal{P}_\varepsilon$ are given by:

$$\mathcal{P}_k = \mu_t \left(\nabla \overline{\boldsymbol{u}} + (\nabla \overline{\boldsymbol{u}})^{\mathrm{T}}\right)^2, \tag{3.62}$$

$$\mathcal{P}_\varepsilon = C_{1\varepsilon} \frac{\varepsilon}{k} \mathcal{P}_k. \tag{3.63}$$

The dissipation of energy is represented by the third term (III), while the last term (IV) accounts for the damping of the turbulent parameters in the mushy/solid region:

$$\boldsymbol{S}_{k,\mathrm{D}} = -\mathcal{A}(\gamma) k, \tag{3.64}$$

$$\boldsymbol{S}_{\varepsilon,\mathrm{D}} = -\mathcal{A}(\gamma) \varepsilon, \tag{3.65}$$

where A is the porosity function and is defined in equation (3.44). In the above equations, the values employed for the model constants $C_{1\varepsilon}$, $C_{2\varepsilon}$, C_μ, σ_k and σ_ε are 1.44, 1.92, 0.09, 1.0 and 1.3, respectively [52].

3.5 Numerical Methods

The previously discussed numerical model is established on the framework of the Finite Volume Method (FVM) based open-source software, OpenFOAM. Consequently, this section aims to provide a basic introduction to FVM within OpenFOAM, encompassing key aspects of the solver implementation and addressing topics such as Multidimensional Universal Limiter for Explicit Solutions (MULES), numerical treatment of the solid-liquid phase change source term, pressure-velocity coupling and the incorporation of an adjustable time-step. Additionally, the solution procedure of the numerical model is detailed, providing a comprehensive understanding of the model implementation within the OpenFOAM environment.

Finite Volume Method

In the FVM, the computational domain is discretised into several control volumes (CVs). A typical CV is illustrated in the Figure 3.2, where P and N represent the owner and neighbouring cells, located at the centroid of the CVs. The cell faces in the computational domain can be classified into two groups: internal faces (between two cells) and boundary faces (at the boundaries of the computational domain). The internal face area vector S_f is normal to the cell face f and points outwards from the owner cell (P) into neighbouring cells (N) at each face of the CV, whereas the face area vectors at the boundary point outwards from the computational domain. Finally, $\boldsymbol{d}$ is the vector joining two points P and N [41, 53].

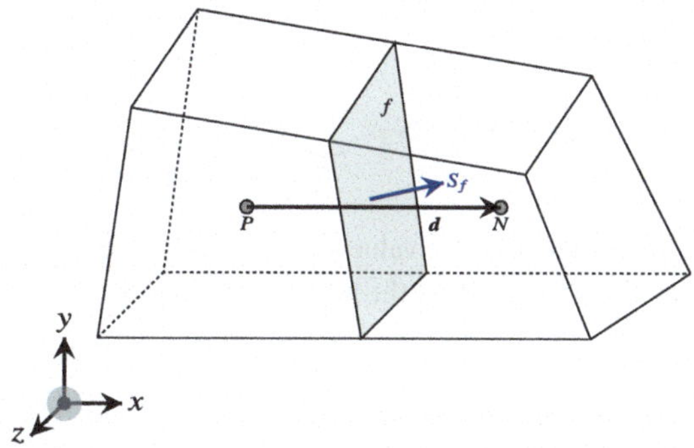

Figure 3.2 Schematic illustration of control volume and finite volume discretisation parameters, adapted from [41, 53]

The discretisation of the above discussed governing equations is explained using the general form of the transport equation for any generic variable φ and is given by [41, 52]

$$\underbrace{\frac{\partial(\varrho\varphi)}{\partial t}}_{\text{temporal term}} + \underbrace{\nabla \cdot (\varrho \boldsymbol{u}\varphi)}_{\text{convective term}} = \underbrace{\nabla \cdot (\Gamma \nabla \varphi)}_{\text{diffusive term}} + \underbrace{S_\varphi}_{\text{source term}}, \tag{3.66}$$

where Γ is the diffusivity associated with the generic variable φ. In the equation (3.66), the temporal term describes the rate of change per unit volume. The convective and diffusive term describes the rate of transport due to convection and diffusion per unit volume, respectively. Finally the source term represents pressure, body force, surface tension force, etc. per unit volume.

The general form of the transport equation (3.66) can be expressed in the integral form over a CV (V_P) at the point P and time interval $[t, t + \Delta t]$:

$$\int_t^{t+\Delta t}\left[\frac{\partial}{\partial t}\int_{V_P}\varrho\varphi dV + \int_{V_P}\nabla\cdot(\varrho\boldsymbol{u}\varphi)dV\right]dt = \int_t^{t+\Delta t}\left[\int_{V_P}\nabla\cdot(\Gamma\nabla\varphi)dV + \int_{V_P}S_\varphi dV\right]dt, \tag{3.67}$$

Discretisation of a general transport equation

Convective term

The discretisation of the convective term in equation (3.67) is performed by transforming the volume integral to surface integral using the Gauss's divergence theorem, and is given by [32, 41, 53]:

$$\int_{V_P}\nabla\cdot(\varrho\boldsymbol{u}\varphi)dV = \int_{S_f}(\varrho\boldsymbol{u}\varphi)\cdot d\boldsymbol{S} \approx \sum_f \boldsymbol{S}_f\cdot(\varrho\boldsymbol{u})_f\varphi_f = \sum_f F\varphi_f, \tag{3.68}$$

where the subscript f denotes the value of the variable at the cell-face and F represents the mass flux ($\boldsymbol{S}_f \cdot (\varrho\boldsymbol{u})_f$) through the surface S_f.

Diffusive term

The discretisation of the diffusive term in equation (3.67) is performed analogously to the convective term, i.e., by transforming the volume integral to surface integral as follows [32, 41, 53]:

$$\int_{V_P}\nabla\cdot(\Gamma\nabla\varphi)dV = \int_{S_f}(\Gamma\nabla\varphi)\cdot d\boldsymbol{S} \approx \sum_f \Gamma_f\boldsymbol{S}_f\cdot(\nabla\varphi)_f. \tag{3.69}$$

In the above equation (3.69), the expression $\boldsymbol{S}_f \cdot (\nabla\varphi)_f$ is treated differently for orthogonal and non-orthogonal grids. Figure 3.3 illustrates the vector representation of a non-orthogonal grid. In case of orthogonal grids, $\boldsymbol{S}_f = \boldsymbol{D}_f$, the expression is approximated as:

$$\boldsymbol{S}_f\cdot(\nabla\varphi)_f = \boldsymbol{D}_f\cdot(\nabla\varphi)_f = |\boldsymbol{D}_f|\frac{\varphi_N - \varphi_P}{|\boldsymbol{d}|}, \tag{3.70}$$

where $|\boldsymbol{d}|$ is the length of the vector PN, whereas φ_P and φ_N are the values of the variable φ at the owner and neighbouring cell, respectively. This approximation is second-order accurate. However, for non-orthogonal grids, the expression is split

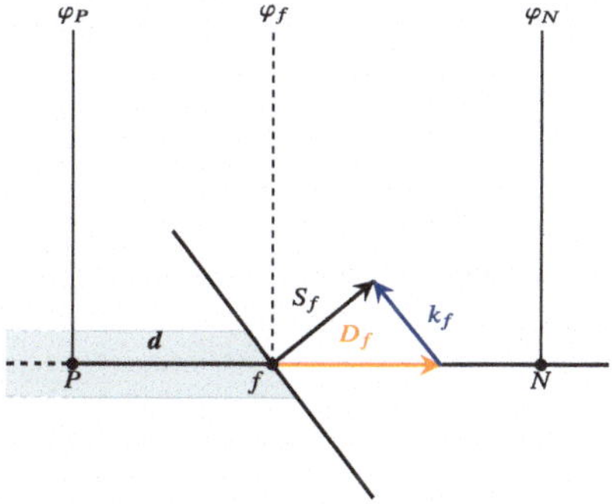

Figure 3.3 Schematic representation of the vectors $\boldsymbol{d}$ and S_f on a non-orthogonal mesh

into orthogonal contribution and non-orthogonal correction, such that $S_f = D_f + k_f$, to preserve second-order accuracy and is given by [41]:

$$S_f \cdot (\nabla\varphi)_f = \underbrace{|D_f|\frac{\varphi_N - \varphi_P}{|d|}}_{\substack{\text{orthogonal}\\ \text{contribution}}} + \underbrace{k_f \cdot (\nabla\varphi)_f}_{\substack{\text{non-orthogonal}\\ \text{correction}}} \tag{3.71}$$

where the vectors Δ and k are determined by the non-orthogonality treatment. A detailed description of various approaches to compensate for the non-orthogonality is discussed by JASAK [53].

Source term

The volume integral of the source term in equation (3.67) can be discretised as the product of the mean value of the integrand ($\overline{S}_\varphi$) and volume V_P of the CV and is written as:

$$\int_{V_P} S_\varphi dV \approx \overline{S}_\varphi V_P \approx S_{\varphi,P} V_P, \tag{3.72}$$

where $S_{\varphi,P}$ is the value of S_φ at the cell center P of the CV. This approximation is second-order accurate [46, 48].

In particular situations, the source term S_φ may be a general function of φ. In such cases the source term is first linearised [52, 54]:

$$S_\varphi(\varphi) = S_u + S_P\varphi, \tag{3.73}$$

where S_u and S_p may also depend on φ. Therefore, the volume integral of the equation (3.73) is given by:

$$\int_{V_P} S_\varphi(\varphi) dV = S_u V_P + S_p \varphi V_P. \tag{3.74}$$

In the above equation (3.74) the term S_u is treated explicitly, whereas the term S_p is most often treated implicitly.

Face interpolation

To discretise the convective (equation (3.68)) and diffusive (equation (3.69)) fluxes, the value of the field variable φ and gradient $\nabla\varphi$ at the cell faces of a CV are required. This is generally approximated by expressing the cell face value in terms of cell centre value. In the present work, a variety of interpolation schemes are used and few of them are discussed below:

- **Upwind Differencing Scheme (UDS)** a.k.a *Upstream Differencing Scheme* or the *donor-cell method* considers the direction of the flow, to determine the value at the cell face [30, 52]. Figure 3.4 shows a schematic representation of the cell centre and cell face values based on positive (blue) and negative (red) flow direction on a one-dimensional (1D) equidistant grid.
  The value of the field variable φ at the cell face *e* is equal to the cell centre value upstream of face *e* and is given by:

$$\varphi_e = \begin{cases} \varphi_P & \text{for } F \geq 0 \\ \varphi_E & \text{for } F < 0. \end{cases} \tag{3.75}$$

  This approximation guarantees boundedness of the solution i.e., it does not cause unrealistic oscillations in the solution. However, it is only first-order accurate and introduces large numerical diffusion.

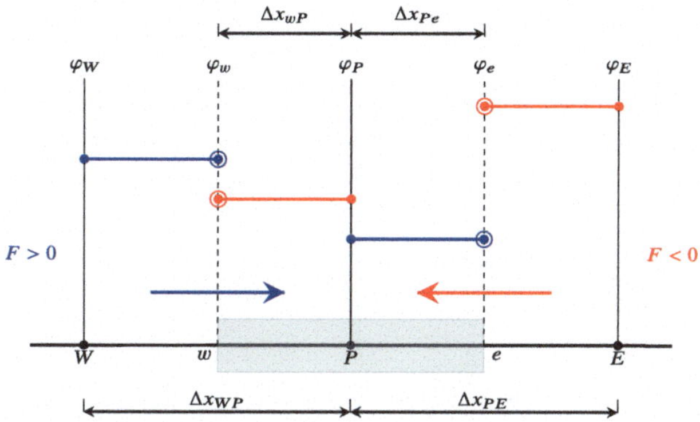

Figure 3.4 Illustration of UDS on 1D equidistant grid

- **Central Differencing Scheme (CDS)** a.k.a *Linear interpolation* is the simplest and straightforward second-order accurate approximation. Figure 3.5 illustrates the cell centre and cell face values on an 1D equidistant grid. The cell face value of the field variable φ at face e is linearly interpolated between the two nearest cell center values:

$$\varphi_e = \varphi_E f_e + \varphi_P (1 - f_e), \tag{3.76}$$

where f_e is the linear interpolation factor and is expressed as:

$$f_e = \frac{\overline{eE}}{\overline{PE}} = \frac{x_E - x_e}{x_E - x_P}. \tag{3.77}$$

Unlike the UDS, this approximation does not consider the direction of the flow and may result in oscillatory solutions [54].

- **Second-order Upwind Differencing Scheme (UDS2)** a.k.a *Linear Upwind Differencing Scheme (LUDS)* is the most popular and commonly used second-order accurate approximation. This scheme is approximated by extrapolating the value of the field variable φ at the cell face e using two upwind cell centre values (see Figure 3.6) and can be estimated as follows:

$$\varphi_e = \begin{cases} \varphi_P + \frac{1}{2}(\varphi_P - \varphi_W) = \frac{1}{2}(3\varphi_P - \varphi_W) & \text{for } F \geq 0 \\ \varphi_E + \frac{1}{2}(\varphi_E - \varphi_{EE}) = \frac{1}{2}(3\varphi_E - \varphi_{EE}) & \text{for } F < 0. \end{cases} \tag{3.78}$$

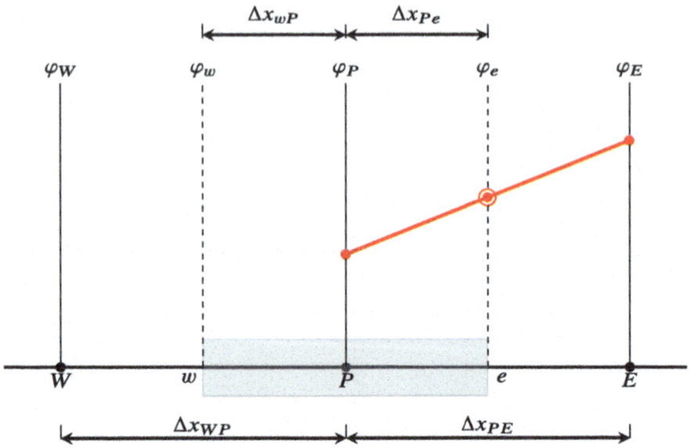

Figure 3.5 Illustration of CDS on 1D equidistant grid

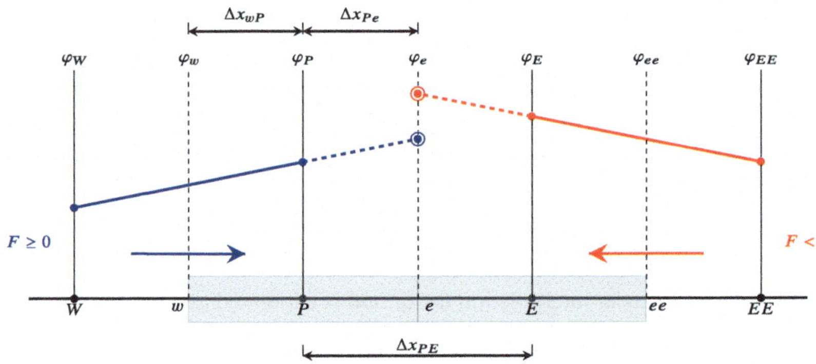

Figure 3.6 Illustration of LUDS on 1D equidistant grid

- **Blended Differencing (BD)** schemes were introduced as an attempt to improve the stability of the scheme (i.e., to avoid unrealistic oscillations in the solution) and to maintain the accuracy of the solution. This is obtained by blending two or more different approximations, the popular one being the combination of UDS (for stability) and CDS (for accuracy). The BD schemes in general can be expressed as:

$$\varphi_e = \zeta (\varphi_e)_{CDS} + (1 - \zeta)(\varphi_e)_{UDS}, \tag{3.79}$$

where $\zeta \in [0, 1]$ is the blending factor to switch between schemes. For $\zeta = 0$ the field variable φ at cell face e is approximated using UDS and for $\zeta = 1$ using

CDS. Various High Resolution Schemes (HRS), Total Variation Diminishing (TVD) and Normalised Variable Approach (NVA) discussed in the literature, e.g. [46, 53, 55–58], can be used to compute the blending factor ζ.

Temporal term

The temporal term in equation (3.67) is discretised using the first-order implicit Euler scheme in the following form:

$$\frac{\partial}{\partial t}\int_{V_P} \varrho\varphi dV \approx \frac{\varrho_P^l \varphi_P^l - \varrho_P^{l-1}\varphi_P^{l-1}}{\Delta t} V_P, \tag{3.80}$$

where φ^l and φ^{l-1} represent the value at the current $(t + \Delta t)$ and previous (t) time step. Furthermore, a second-order accurate implicit Euler scheme (backward differencing scheme) can also be used for the discretisation of the temporal term and is given by:

$$\frac{\partial}{\partial t}\int_{V_P} \varrho\varphi dV \approx \frac{\frac{3}{2}\varrho_P^l \varphi_P^l - 2\varrho_P^{l-1}\varphi_P^{l-1} + \frac{1}{2}\varrho_P^{l-2}\varphi_P^{l-2}}{\Delta t} V_P, \tag{3.81}$$

where φ^{l-2} represents the value at time $(t - \Delta t)$. The first-order Euler scheme is unconditionally stable and is used in majority of the cases discussed in this work.

Numerical treatment of the VoF method

The semi-discretised form of equation (3.24) is given by:

$$\frac{\partial \alpha_1}{\partial t} + \sum_f S_f \cdot (\alpha_1 \boldsymbol{u})_f + \sum_f S_f \cdot [\alpha_1(1-\alpha_1)\boldsymbol{u}_{\mathbf{r}}]_f = 0. \tag{3.82}$$

The temporal term in the above equation is now discretised using first-order explicit Euler scheme and the remaining terms can be re-written in terms of face volumetric fluxes, which reads [59]:

$$\frac{\alpha_1^{l+m+1} - \alpha_1^l}{\Delta t} V_P + \sum_f \left[(\alpha_1^{l+m})_f \mathcal{F}^{L,l+m} + \left(\alpha_1^{l+m}(1-\alpha_1^{l+m})\right)_f \mathcal{F}_{\mathrm{r}}^{L,l+m} \right] = 0, \tag{3.83}$$

where $\mathcal{F} = \boldsymbol{u}_f \cdot \boldsymbol{S}_f$ is the face volumetric flux, m the number of actual fixed-point inner iteration in order to handle the non-linearity of the fluxes and $\mathcal{F}_\mathrm{r}$ the face volumetric flux based on relative compression velocity $\boldsymbol{u}_\mathbf{r}$ and is defined as:

$$\mathcal{F}_\mathrm{r} = n_f \min\left[C_\alpha \frac{|\mathcal{F}^{L,l}|}{|S_f|}, \max\left(\frac{|\mathcal{F}^{L,l}|}{|S_f|}\right)\right]. \tag{3.84}$$

Here, C_α is the compression coefficient, which determines the strength of the interface compression. The value $C_\alpha = 0$ represents no compression and $C_\alpha = 1$ corresponds to conservative compression. The interface unit normal flux n_f is defined as:

$$n_f = \frac{(\nabla\alpha_1)_f}{|(\nabla\alpha_1) + \delta_n|} \cdot S_f, \tag{3.85}$$

where δ_n is a stabilisation factor to avoid "division by zero" error and is expressed as:

$$\delta_n = \frac{\varepsilon}{[(\sum_N V_i)/N]^{1/3}}, \tag{3.86}$$

with N being the number of computational cells and ε a small number.

The face values for volume fraction α_1 are calculated using BD schemes [equation (3.79)]. In this study, the blending factor ζ is computed using the following expression [59, 60]:

$$\zeta = \min\left\{\max\left(1 - \max\left[\{1 - 4\alpha_{1,P}\{1 - \alpha_{1,P}\}\}^2, \{1 - 4\alpha_{1,N}\{1 - \alpha_{1,N}\}\}^2\right], 0\right), 1\right\}. \tag{3.87}$$

Numerical treatment of the solid-liquid phase change source term

In this study, the energy conservation equation is solved using a source based method [44, 61, 62]. The non-linear aspect of the energy conservation equation (3.40) is addressed by considering it as a source term (S_h). As a result, both temperature T and melt fraction γ, where γ is a function of temperature, are involved in a single equation. Due to the interdependence of these variables, an iterative procedure is required to achieve consistency between the two fields and to ensure the overall satisfaction of energy conservation equation. The iterative procedure in a single time-step can be described as follows:

Step 1: At the beginning of the iteration $\gamma^{i} = \gamma^{\text{old}}$, where γ^{old} is the melt fraction field obtained at the end of the preceding time-step.

Step 2: The equation (3.88) is solved to determine the temperature field T^{i+1} with γ^{i}, where $i+1$ represents the current iterative step, while i corresponds to the previous iterative step:

$$c_p \frac{\partial(\varrho T^{i+1})}{\partial t} + c_p \nabla \cdot (\varrho \boldsymbol{u} T^{i+1}) = \nabla \cdot (\kappa \nabla T^{i+1}) \underbrace{- \alpha_1 \mathcal{L} \left[\frac{\partial \varrho \gamma^{i}}{\partial t} + \nabla \cdot (\varrho \boldsymbol{u} \gamma^{i}) \right]}_{S_h}. \tag{3.88}$$

Step 3: The melt fraction field is updated based on the new temperature field T^{i+1}:

$$\gamma^{i+1,*} = \gamma^{i} + \omega \frac{c_p}{\mathcal{L}} (T^{i+1} - T^{\text{corr}}), \tag{3.89}$$

where ω is the under-relaxation factor, while T^{corr} is the correction temperature, which is calculated utilising the melt fraction from the preceding iteration γ^{i} and is give by:

$$T^{\text{corr}} = (T_{\text{l}} - T_{\text{s}})\gamma^{i} + T_{\text{s}}. \tag{3.90}$$

Step 4: During the iterative process, it is possible that the calculated melt fraction $\gamma^{i+1,*}$ may result in non-physical values, such as a value greater than one or less than zero. To prevent this and ensure boundedness, the melt fraction values are corrected across the computational domain using the following expression:

$$\gamma^{i+1} = \min\left\{ \max\{\gamma^{i+1,*}, 0\}, 1 \right\}. \tag{3.91}$$

Step 5: The material properties are updated based on equation (3.41) in accordance with the newly calculated melt fraction (γ^{i+1}):

$$\phi = \alpha_1 \big[\underbrace{\gamma^{i+1} \phi_{\text{l}} + (1 - \gamma^{i+1}) \phi_{\text{s}}}_{\phi_1} \big] + \alpha_2 \phi_2, \tag{3.92}$$

Step 6: **Steps 2-5** are repeated until convergence is achieved, which is determined by the convergence criterion ϵ_T and reads:

$$|T_P^{i+1} - T_P^{i}| < \epsilon_T. \tag{3.93}$$

System of linear equations

The integral form of the transport equation (3.67) is converted into a linear algebraic equation at each cell centre of the CV using the above discussed finite volume discretisation and face interpolation schemes [46, 52]:

$$a_P \varphi_P + \sum_N a_N \varphi_N = b, \tag{3.94}$$

where a_P and a_N are coefficients of the generic field variable φ at P and its neighbour N, respectively. The system of equations for the computational domain is expressed in a matrix form:

$$[A]\{\varphi\} = \{b\}, \tag{3.95}$$

where $[A]$ represents a sparse matrix with a_P as diagonal coefficients and a_N as off-diagonal coefficients, $\{\varphi\}$ the unknowns of the system, and $\{b\}$ the array of source terms.

The solution of equation (3.95) is obtained using iterative methods i.e., starting with an initial guess and subsequently improving the approximation until convergence criteria is achieved. In this thesis, both the Preconditioned Conjugate Gradient (PCG) with Geometric agglomerated Algebraic Multigrid (GAMG) as preconditioner for symmetric matrices and the Preconditioned Bi-Conjugate Gradient Stabilised (PBiCGStab) with Diagonal-based Incomplete Lower-Upper (DILU) as preconditioner for asymmetric matrices have been used (refer to Ferziger and Perić [46], Moukalled et al. [48], and Saad [63] for further information on these methods).

Pressure-velocity coupling

One of the source term in the momentum equation (3.2) is the pressure gradient, thus, it is obvious that the pressure distribution in the computational domain is needed, to solve the momentum equation. Also, a correct pressure field is required for the resulting velocity field to satisfy the continuity equation [52, 54]. However, unlike the other field variables, i.e., velocity, temperature, volume fraction, etc., pressure does not have a dedicated transport equation to be calculated directly. Therefore, the pressure equation is generally derived from conservation of mass and a semi-discretised momentum equation. Following equation (3.94), the semi-discretised momentum equation can be expressed as:

$$a_\text{P}\boldsymbol{u}_\text{P} + \sum_\text{N} a_\text{N}\boldsymbol{u}_\text{N} = b - \nabla p, \tag{3.96}$$

$$a_\text{P}\boldsymbol{u}_\text{P} = H(\boldsymbol{u}) - \nabla p \tag{3.97}$$

In the above equations, the pressure gradient is still in its differential form [64] (i.e., it is not discretised) for the sake of derivation and the term $H(\boldsymbol{u})$ accounts for the off-diagonal coefficient matrix of the velocity field, and all the source terms, including the source part of the transient terms [53]:

$$H(\boldsymbol{u}) = -\sum_\text{N} a_\text{N}\boldsymbol{u}_\text{N} + b. \tag{3.98}$$

Using equation (3.97) the velocity at the cell centre reads:

$$\boldsymbol{u}_\text{P} = (a_\text{P})^{-1}[H(\boldsymbol{u}) - \nabla p], \tag{3.99}$$

and the velocity at the cell face can be interpolated as:

$$\boldsymbol{u}_f = (a_\text{P})_f^{-1}[(H(\boldsymbol{u}))_f - (\nabla p)_f], \tag{3.100}$$

Following equation (3.68), the incompressible continuity equation (3.5) can be discretised as:

$$\nabla \cdot \boldsymbol{u} = \sum_f \boldsymbol{u}_f \cdot S_f = 0. \tag{3.101}$$

Substituting equation (3.100) in equation (3.101), the pressure equation for incompressible fluid flow is obtained:

$$\nabla \cdot [(a_\text{P})^{-1}\nabla p] = \nabla \cdot [(a_\text{P})^{-1} H(\boldsymbol{u})]. \tag{3.102}$$

Therefore, the final set of discrete equations for the Navier-Stokes system is given by:

$$a_\text{P}\boldsymbol{u}_\text{P} = H(\boldsymbol{u}) - \sum_f S_f(p)_f, \tag{3.103}$$

$$\sum_f S_f[(a_\text{P})_f^{-1}(\nabla p)_f] = \sum_f S_f[(a_\text{P})_f^{-1}(H(\boldsymbol{u}))_f]. \tag{3.104}$$

Subsequently, the face flux is computed using equation (3.100) and reads:

$$\mathcal{F}_f = S_f \cdot \boldsymbol{u}_f = S_f \cdot \left\{ (a_{\mathrm{P}})_f^{-1} [(H(\boldsymbol{u}))_f - (\nabla p)_f] \right\}. \tag{3.105}$$

Numerous algorithms are proposed to achieve the coupling of velocity with the pressure gradient. The most commonly used algorithms are **S**emi-**I**mplicit **M**ethod for **P**ressure **L**inked **E**quations (SIMPLE) for steady state flows proposed by Patankar and Spalding [65] and **P**ressure **I**mplicit with **S**plitting of **O**perators (PISO) for transient flows proposed by Issa [66]. A detailed description of these algorithms can also be found in Nikrityuk [30], Ferziger and Perić [46], Versteeg and Malalasekera [52].

Adjustable time-step

To ensure the stability of the computational process a well-accepted time-step control defined by Courant-Friedrichs-Lewy (CFL) is utilised. The CFL condition relates the spatio-temporal discretisation of the transport equation. For one-dimensional numerical problems, the CFL condition is given by [67]:

$$Co = \frac{|\boldsymbol{u}|\Delta t}{\Delta x}, \tag{3.106}$$

where Co is the Courant number, $|\boldsymbol{u}|$ the velocity magnitude, Δt the time-step and Δx the cell size. In other words, the CFL condition dictates how much information ($\boldsymbol{u}$) travels across a computational grid cell (Δx) in a given time step (Δt). The Courant number at a cell face is defined as [41]:

$$Co = \frac{|\boldsymbol{u}_f \cdot S_f|}{d \cdot S_f} \Delta t. \tag{3.107}$$

In the simulations, equation (3.107) is used to compute the local Courant number Co_{o} based on the values of $\boldsymbol{u}_f$ and Δt calculated in the previous time-step. The new time-step is computed from the local Courant number by defining a maximum Courant number $Co_{\max}$ and time-step $t_{\max}$, using the following expression [68]:

$$\Delta t^l = \min\left\{ \frac{Co_{\max}}{Co_{\mathrm{o}}} \Delta t^{\mathrm{o}}, \left(1 + \lambda_1 \frac{Co_{\max}}{Co_{\mathrm{o}}}\right) \Delta t^{\mathrm{o}}, \lambda_2 \Delta t^{\mathrm{o}}, \Delta t_{\max} \right\}. \tag{3.108}$$

In the above equation, $Co_{\max}$ and $t_{\max}$ are predefined limit values. If the calculated Co_{o} is greater than the predefined $Co_{\max}$ the time-step will decrease and vice versa [68]. This results in time-step oscillations and may lead to numerical instability in simulations. Therefore, the relaxation factors λ_1 and λ_2 are introduced to avoid such oscillations by damping the increased time-step, and the values are set to 0.1 and 1.2, respectively.

Generally at the start-up of a simulation, the value for the initial time-step Δt_{init} is set to a very small value and the new computed time-step could be too large for the start even with the introduced relaxation factors. In this case, an intermediate time-step value Δt^*_{init} for the initial time-step is evaluated from:

$$\Delta t^*_{\text{init}} = \min\left(\frac{Co_{\text{max}}\Delta t_{\text{init}}}{Co_{\text{o}}}, \Delta t_{\text{max}}\right). \tag{3.109}$$

Therefore, at the beginning of the simulation, the old time-step Δt_{o} in equation (3.108) is replaced with the intermediate time-step value Δt^*_{init}, such that, the resulting local Courant number Co_{o} is close to the predefined maximum Courant number Co_{max}. In this study, the Co_{max} value is set to less than 0.2 for mould filling and less than 0.75 for solidification simulations.

Solution algorithm

The solution algorithm comprises the following steps to predict the dynamics of 3D multi-fluid, multiphase heat transfer and solid-liquid phase change during the mould filling and cooling stages of the AT welding process:

Step 1: Initialise all field variables, including $\boldsymbol{u}$, p, T and γ.
Step 2: Adjust the time step Δt based on the maximum Courant number using equation (3.108)
Step 3: m Calculate the volume fraction α (equation (3.24)) using MULES algorithm, as described by equations (3.83)–(3.86).
Step 4: Solve the energy equation (3.88) to account for the heat transfer in the system.
Step 5: Calculate the melt fraction γ based on the temperature field T using equations (3.89)–(3.92).
Step 6: Repeat **Step 4** and **Step 5** iteratively until convergence is reached, as determined by equation (3.93).
Step 7: Solve the pressure-velocity coupling involving equations (3.98)–(3.105). The PISO algorithm [66] is utilised for this purpose.
Step 8: If required, solve turbulence equations, i.e., equations (3.57) and (3.58) to account for turbulent effects in the flow.
Step 9: Update material properties (ϱ, μ, c_p, κ) using equation (3.41).
Step 10: Advance to the next time step and repeat the process from **Step 2** or finish the calculation if the desired simulation time is reached.

Figure 3.7 illustrates an overview of the entire solution algorithm in the form of a flow chart.

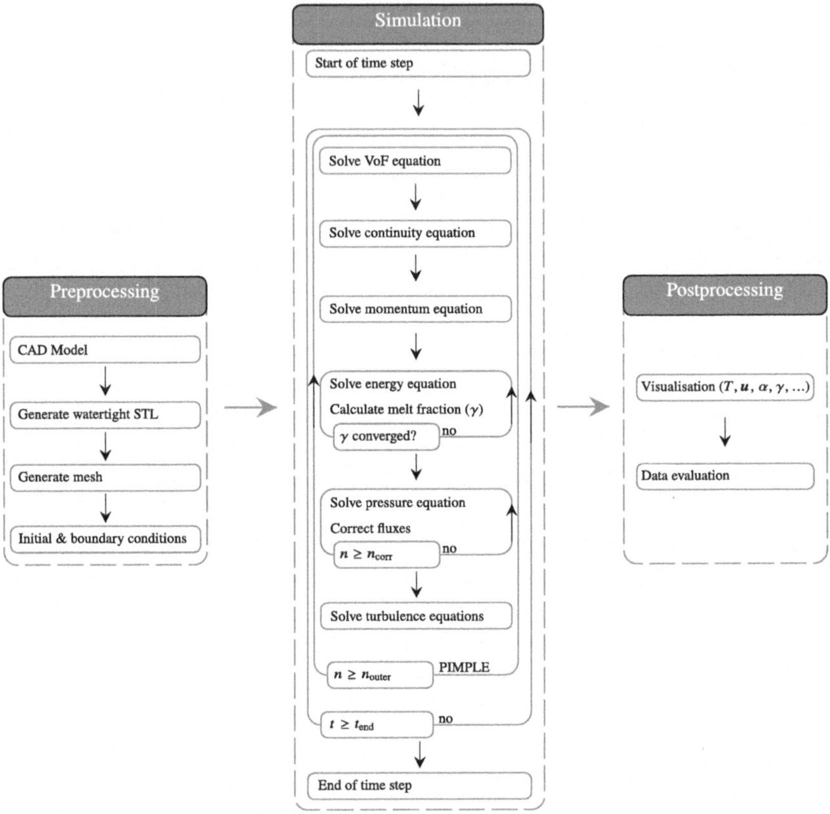

Figure 3.7 Graphical representation of the developed numerical model for predicting the dynamics of 3D multi-fluid, multiphase heat transfer and solid-liquid phase change throughout the mould filling and cooling stages of the AT welding process

Note: For PISO, n_{outer} is set to 1.

3.6 OpenFOAM

OpenFOAM (Open-source **F**ield **O**peration **A**nd **M**anipulation) is an open-source, object-oriented library, Finite Volume Method (FVM) based Computational Fluid Dynamics (CFD) toolbox written in C++ programming language to solve systems of linked partial differential equations (PDE). It was originally developed as an in-house code FOAM [32, 41, 53, 69] at the IMPERIAL COLLEGE LONDON and was first released to public as a commercial code. In 2004, the code was finally released as open-source under the GNU General Public Licence (GPL). The development of OpenFOAM is currently active and it has a very strong user base in both industry and academia.

With the fundamental concepts in Object-Oriented Programming (OOP), such as, encapsulation, inheritance, abstraction, polymorphism, etc., the OpenFOAM code interface is so developed, that the PDEs required to solve the complex physical phenomena in continuum mechanics are expressed in a form of the mathematical nomenclature [69, 70] as we know it. For example, the OpenFOAM implementation of the energy equation (3.88) is given by:

```
1  ...
2  fvVectorMatrix TEqn
3  (
4     cp*fvm::ddt(rho, T)
5  + cp*fvm::div(rhoPhi, T)
6  + alpha1*L*(fvc::ddt(rho,gamma) + fvc::div(rhoPhi,gamma))
7  - fvm::laplacian(fvc::interpolate(kappa), T)
8  );
9  TEqn.relax()
10 TEqn.solve()
11 ...
```

Here `ddt`, `div`, `laplacian` and `grad` are the discrete differential operators. The namespace `fvm` (finite volume matrix) evaluates an operator implicitly and returns a coefficient matrix, whereas the namespace `fvc` (finite volume calculus) evaluates an operator explicitly and returns the actual field values. The result of mathematical operations of the discrete differential operators acting upon the ingredients `rho, cp, T, phi, gamma` and `kappa` (from line no. 4–7) is a coefficient matrix of type `fvVectorMatrix`.

4 Validation

This chapter is dedicated to validating the aforementioned numerical model for the prediction of heat transfer and solid-liquid phase change. The validation process is initiated by exploring a series of well-documented reference cases. These reference cases are selected to progressively increase the complexity of the scenarios under investigation. Firstly, a simple one-dimensional (1D) two-phase Stefan problem is considered. Next, moving from 1D to 2D, the melting of pure gallium in a rectangular enclosure is investigated. Finally, the complexity is further raised by investigating a multi-fluid system featuring free surface thermocapillary flow with phase change. Additionally, a low-cost, in-house experiment employing a transparent phase change material (PCM), specifically paraffin wax (RubiTherm® 31), is designed, fabricated and subsequently, the experimental results are used for validation, to provide a comprehensive evaluation of the developed numerical model. It is noteworthy that partial results outlined in this chapter have been previously published in Kewalramani et al. [71].

4.1 One-dimensional two-phase Stefan Problem

4.1.1 Problem Description

The one-dimensional heat conduction controlled melting of a semi-infinite slab as depicted in Figure 4.1, is initially in the solid state, at a uniform temperature $T_{\mathrm{ini}} = T_{\mathrm{c}}$ (which is less than the melting temperature T_{m}), is considered. The melting is initiated by imposing a constant temperature T_{h} (which is greater than the melting temperature T_{m}) on the left boundary i.e., at $x = 0$. As a consequence, the slab starts to melt and the solid-liquid interface front progresses towards the right. The location of the solid-liquid interface is denoted as $x_{\mathrm{m}}(t)$. Therefore, in this problem the heat

R. G. Kewalramani, *Computational Thermo-Fluid Dynamics of Aluminothermic Welding Process*, https://doi.org/10.1007/978-3-658-47535-2_4

conduction equation is solved in both phases with constant (Dirichlet) boundary conditions at the left and right boundaries and the Stefan condition at the solid-liquid interface. The governing equations for each phase are as follows:

$$\frac{\partial T}{\partial t} = a_{\mathrm{l}} \frac{\partial^2 T}{\partial x^2} \qquad 0 \leq x \leq x_{\mathrm{m}}(t), \tag{4.1}$$

$$\frac{\partial T}{\partial t} = a_{\mathrm{s}} \frac{\partial^2 T}{\partial x^2} \qquad x_{\mathrm{m}}(t) < x < \infty, \tag{4.2}$$

where a_{l} and a_{s} are the thermal diffusivity of the material in liquid and solid phases, respectively, and is computed for each phase as follows:

$$a_{\mathrm{l}} = \frac{\kappa_{\mathrm{l}}}{\varrho_{\mathrm{l}}\, c_{p,\mathrm{l}}}, \qquad a_{\mathrm{s}} = \frac{\kappa_{\mathrm{s}}}{\varrho_{\mathrm{s}}\, c_{p,\mathrm{s}}} \tag{4.3}$$

Herein, it is assumed that all the material properties in each phase are constant, moreover, there is no fluid flow in the melt [30, 72]. The initial and boundary conditions for the liquid phase are given as:

$$T(x = 0, t) = T_{\mathrm{h}}; \qquad \text{for } t > 0, \tag{4.4}$$

$$T(x_{\mathrm{m}}(t), t) = T_{\mathrm{m}}; \qquad \text{for } t > 0, \tag{4.5}$$

$$T(x, t = 0) = T_{\mathrm{ini}}; \qquad \text{for } x > 0. \tag{4.6}$$

The initial and boundary conditions for the solid phase are given as:

$$T(x_{\mathrm{m}}(t), t) = T_{\mathrm{m}}; \qquad \text{for } t > 0, \tag{4.7}$$

$$T(x \to \infty, t) = T_{\mathrm{c}}; \qquad \text{for } x > 0. \tag{4.8}$$

The boundary condition at the solid-liquid interface $x_{\mathrm{m}}(t)$ is given as:

$$\varrho \mathcal{L} \frac{dx_{\mathrm{m}}}{dt} = \kappa_{\mathrm{s}} \frac{\partial T}{\partial x} - \kappa_{\mathrm{l}} \frac{\partial T}{\partial x}, \tag{4.9}$$

where $\varrho = \varrho_{\mathrm{l}} = \varrho_{\mathrm{s}}$ and is known as the Stefan condition.

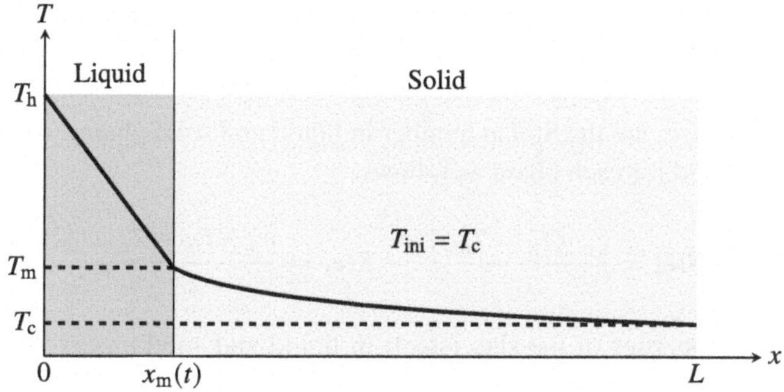

Figure 4.1 Illustration of the 1D two-phase Stefan problem

Analytical solution

The analytical solutions for the temperature distribution in the solid and liquid phases are obtained by solving the equations (4.1) and (4.2) with the given initial and boundary conditions i.e., equations (4.4)–(4.8) can be computed as follows:

$$T(x,t) = \begin{cases} T_{\mathrm{h}} - \dfrac{(T_{\mathrm{h}} - T_{\mathrm{m}})\,\mathrm{erf}\left(\dfrac{x}{2\sqrt{a_{\mathrm{l}}t}}\right)}{\mathrm{erf}\,(\eta)} & 0 \le x \le x_{\mathrm{m}}(t), \\ T_{\mathrm{c}} + \dfrac{(T_{\mathrm{m}} - T_{\mathrm{c}})\,\mathrm{erfc}\left(\dfrac{x}{2\sqrt{a_{\mathrm{s}}t}}\right)}{\mathrm{erfc}\left(\eta\sqrt{a_{\mathrm{l}}/a_{\mathrm{s}}}\right)} & x > x_{\mathrm{m}}(t), \end{cases} \tag{4.10}$$

where erf (ψ) [1] is the error function and is defined as:

$$\mathrm{erf}\,(\psi) = \frac{2}{\sqrt{\pi}} \int_0^x \exp\,(-\psi^2)d\psi, \tag{4.11}$$

while erfc (ψ) is the complimentary error function and can be calculated as: erfc $(\psi) = 1 - \mathrm{erf}\,(\psi)$. The location of the moving interface is obtained by:

$$x_{\mathrm{m}}(t) = 2\eta\sqrt{a_{\mathrm{l}}t}. \tag{4.12}$$

In the above equations, η is the non-linear eigenvalue and is determined by solving the transcendental equation (4.13) using iterative methods:

[1] Note: erf (0) = 0 and erf $(\pm\infty) = \pm 1$

$$\eta\sqrt{\pi} = \frac{Ste_{\mathrm{l}} \exp\left(-\eta^2\right)}{\operatorname{erf}(\eta)} - \frac{Ste_{\mathrm{s}} \exp\left(-\eta^2 (a_{\mathrm{l}}/a_{\mathrm{s}})\right)}{\sqrt{a_{\mathrm{l}}/a_{\mathrm{s}}} \operatorname{erfc}\left(\eta\sqrt{a_{\mathrm{l}}/a_{\mathrm{s}}}\right)}, \tag{4.13}$$

where Ste_{l} and Ste_{s} are the Stefan number in liquid and solid phases, respectively, and can be defined for each phase as follows:

$$Ste_{\mathrm{l}} = \frac{c_{p,\mathrm{l}}(T_{\mathrm{h}} - T_{\mathrm{m}})}{\mathcal{L}}, \qquad Ste_{\mathrm{s}} = \frac{c_{p,\mathrm{s}}(T_{\mathrm{m}} - T_{\mathrm{c}})}{\mathcal{L}}. \tag{4.14}$$

The material properties of the slab (steel) in liquid and solid phases are given in Table 4.1.

Table 4.1 Material properties of steel in solid and liquid phases

Property	Symbol	Value	Unit
Density, solid	ϱ_{s}	6900	kg m^{-3}
Density, liquid	ϱ_{l}	6900	kg m^{-3}
Specific heat capacity, solid	$c_{p,\mathrm{s}}$	640	J $\mathrm{kg}^{-1}\,\mathrm{K}^{-1}$
Specific heat capacity, liquid	$c_{p,\mathrm{l}}$	680	J $\mathrm{kg}^{-1}\,\mathrm{K}^{-1}$
Thermal conductivity, solid	κ_{s}	30	W $\mathrm{m}^{-1}\,\mathrm{K}^{-1}$
Thermal conductivity, liquid	κ_{l}	55	W $\mathrm{m}^{-1}\,\mathrm{K}^{-1}$
Latent heat of fusion	$\mathcal{L}$	277	kJ kg^{-1}

In this study, the 1D two-phase Stefan problem is investigated for three different combinations of Stefan numbers in liquid and solid phases. The eigenvalues essential for determining the location of the solid-liquid interface front analytically for these combinations, are computed by employing equation (4.13) and the computed values are given in Table 4.2.

Table 4.2 Computed eigenvalues and Stefan numbers for different temperatures imposed on the hot and cold walls

Parameter	Combination			Unit
	1	2	3	
T_{h}	1700	1800	2000	K
T_{c}	1695.25	1690	1680	K
Ste_{l}	0.0110	0.2565	0.7475	–
Ste_{s}	0.0005	0.0127	0.3581	–
η	0.0740	0.3402	0.5393	–

Simulation set-up

A computational domain of size 500 mm $\times$ 1.5 mm (i.e., $L >> H$) is selected and discretised with equidistant cell size of $\Delta x = 1.5$ mm in the $x-$direction, while only one cell is added in the $y-$ and $z-$directions. The top and bottom boundaries are kept insulated. An `empty`[2] boundary condition is imposed on the front and back boundaries of the domain. The melting temperature of the employed material is $T_{\mathrm{m}} = 1695.5$ K. At the right boundary, a constant temperature T_{c} ($< T_{\mathrm{m}}$) is maintained, while the opposite (left) boundary is set to T_{h} ($> T_{\mathrm{m}}$), in accordance with the specified combinations detailed in Table 4.2. At time $t = 0$ s, the slab is at an initial temperature $T_{\mathrm{ini}} = T_{\mathrm{c}}$. Here the solidus temperature is set to $T_{\mathrm{s}} = T_{\mathrm{m}} -$ 0.25 K and the liquidus temperature is set to $T_{\mathrm{l}} = T_{\mathrm{m}} +$ 0.25 K.[3]

4.1.2 Results and Discussions

Figure 4.2 compares the locations of the solid-liquid interface front for three different combinations of Stefan numbers in solid and liquid phases, both analytically and through simulation. The quantitative analysis of the same is provided in Table 4.3, which shows that the simulation values exhibit a relative deviation of less than 0.2 %. from the analytical solutions, indicating a high accuracy of the proposed numerical model.

Additionally, a comparison between the analytical solution (equation (4.10)) and the simulation for the temperature distribution in the slab, specifically for the Stefan numbers combination 2, is illustrated in Figure 4.3. It is noteworthy that a melting range is required to define a solid-liquid interface by the proposed model, unlike the analytical solutions that are derived under the assumption of sharp interface. Despite this distinction, the simulation results consistently demonstrate an excellent agreement with the analytical solution.

---

[2] A special boundary condition in OpenFOAM for cases with reduced dimensions

[3] The developed numerical model cannot handle solidification/melting for a sharp interface (i.e., at a single melting temperature), therefore, a defined melting temperature range of $\Delta T = T_{\mathrm{l}} - T_{\mathrm{s}} = 0.5$ K is employed.

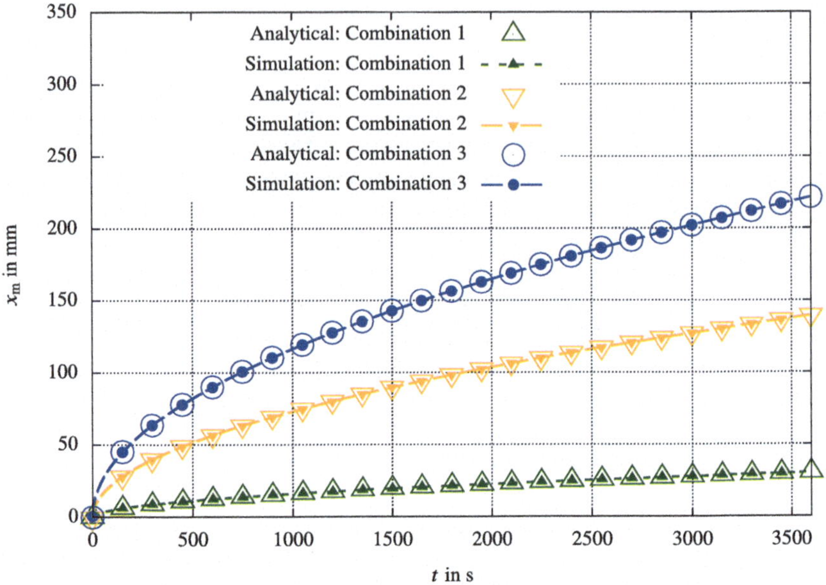

Figure 4.2 Locations of the solid-liquid interface front for three different combinations of Stefan numbers in solid and liquid phases: Analytical versus Simulation

Table 4.3 Quantitative analysis of the locations of the interface front for 3 different combinations of Stefan numbers: Analytical versus Simulation – **Note:** A, S and RE denote Analytical, Simulation and Relative Error, respectively

Location		Combination			Unit
$x_m(t)$		1	2	3	
$x_m(1500)$	**A**	19.637	90.240	143.029	mm
	S	19.613	90.172	142.929	mm
	RE	0.122	0.075	0.069	%
$x_m(3000)$	**A**	27.771	127.619	202.274	mm
	S	27.774	127.582	202.216	mm
	RE	0.01	0.028	0.028	%

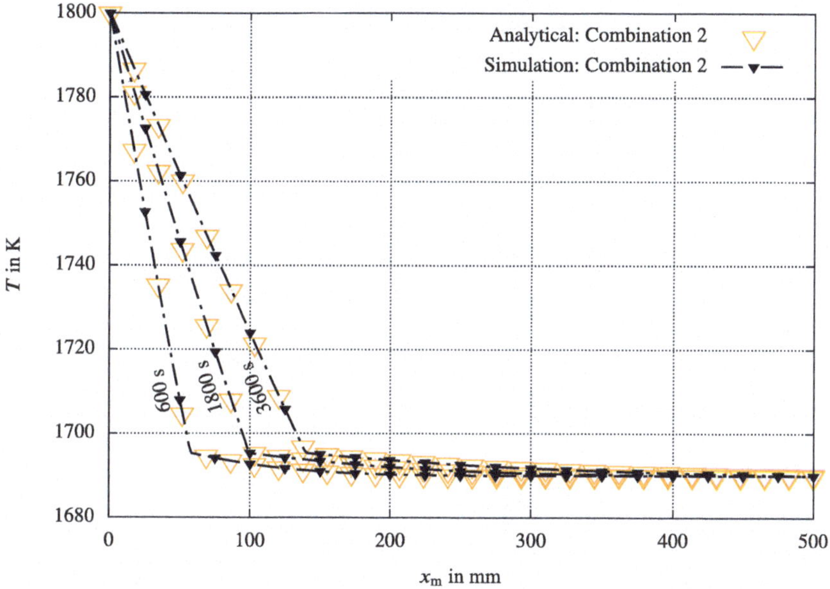

Figure 4.3 Temperature distribution in the slab at three different time instances for the combination 2: Analytical versus Simulation

4.2 Melting of Pure Gallium in a Rectangular Enclosure

4.2.1 Problem Description

The melting of pure gallium (99.6 %) in a rectangular enclosure, as explored in experiments conducted by Gau and Viskanta [73], serves as the classical benchmark case used by various authors, e.g. [24, 43, 74–79], for validating algorithms related to natural convection dominated solid-liquid phase change. The rectangular enclosure has inside dimensions of $L = 88.9$ mm, $H = 63.5$ mm and $W = 38.1$ mm. The left wall is maintained at a constant temperature $T_h = 311.15$ K ($> T_m$), while the right wall is kept at $T_c = 301.45$ K ($< T_m$). The top and bottom walls are assumed to be adiabatic. At time $t = 0$, the gallium inside the enclosure is in the solid state with an initial temperature of $T_{ini} = T_c$. The melting temperature of pure gallium is $T_m = 302.93$ K. The schematic of the enclosure along with the boundary conditions is illustrated in Figure 4.4 and the material properties, used for the simulations, are listed in Table 4.4.

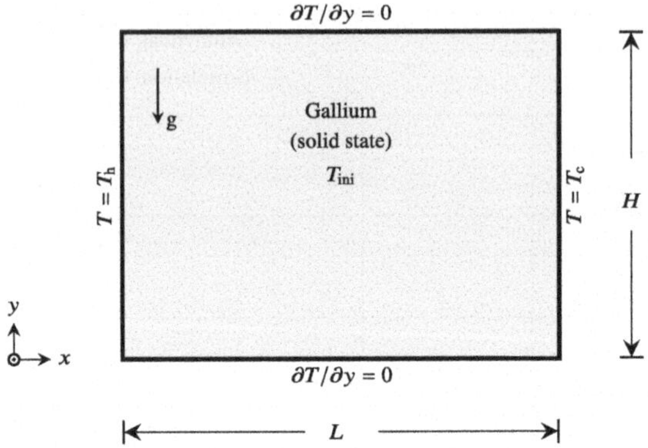

Figure 4.4 Numerical set-up of gallium melting test case

Table 4.4 Material properties of gallium

Property	Symbol	Value	Unit
Density	ϱ	6093	kg m^{-3}
Dynamic viscosity	μ	1.81×10^{-3}	kg m^{-1} s^{-1}
Specific heat capacity	c_p	381.5	J kg^{-1} K^{-1}
Thermal conductivity	κ	32	W m^{-1} K^{-1}
Latent heat of fusion	$\mathcal{L}$	80.160	kJ kg^{-1}
Thermal expansion coeff.	β	1.2×10^{-4}	K^{-1}

4.2.2 Results and Discussions

BRENT et al. [43] were the first to investigate this case numerically. Considering the limited computational resources at the time, a two-dimensional (2D) domain with a grid size of 42×32 hexahedral cells was considered. For the approximation of the convective fluxes, they utilised a first order, UDS. Thus, this set-up serves as the starting point for the present study. Following Brent et al. [43], the mushy zone constant C and the small computational constant ϵ are set to 1.6×10^6 kg m^{-3} s^{-1} and 1×10^{-3}, respectively.

In this study, 2D simulations are conducted with only one cell is added in the z−direction. The front and back boundaries of the enclosure are imposed with an `empty`[2] boundary condition, while the top and bottom boundaries are assumed to be insulated. A grid independent study is performed with three different grid configurations in $x-$ and $y-$directions: 84 $\times$ 64 cells, 126 $\times$ 96 cells and 168 $\times$ 128 cells. The locations of the solid-liquid interface for different grid configurations employed in this study are shown in Figure 4.5. The results show that the deviations in the solid-liquid interface fronts from grid 84 $\times$ 64 are minimal. However, considering both solution accuracy and the available computational resources, the computational grid with 126 $\times$ 96 is chosen and utilised for further investigations.

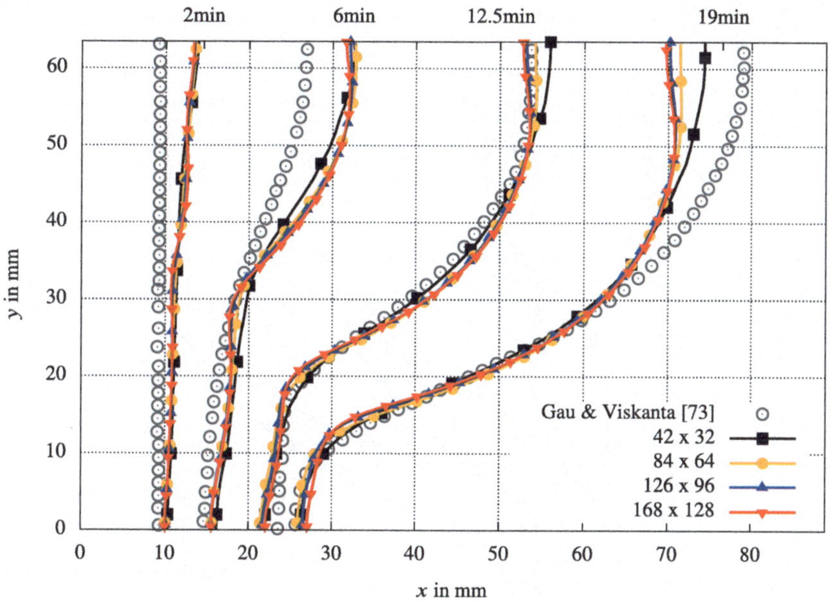

Figure 4.5 Comparison of the solid-liquid interface fronts from experiments [73] and 2D simulations for various grid configurations at different times

Although the first-order UDS (`upwind`) is numerically stable, it is prone to large numerical diffusion. To avoid this numerical diffusion very fine grid configurations are required in order to obtain an accurate solution, which is computationally expensive. Therefore, two other schemes, *viz,* CDS (`linear`) and LUDS (`linearUpwind`) are employed to investigate the influence of the numerical schemes on the temporal evolution of the solid-liquid interface front.

As shown in Figure 4.6, the multi-cellular feature is immediately observed during the initial stages of the melting process when second-order schemes are employed. Initially, the strength of the convective rolls is weak. As the melting progresses, these convective rolls merge, resulting in fewer rolls with a strong velocity field, thereby, enhancing the heat transfer at the solid-liquid interface front leading to such curvy patterns. It is noteworthy that the multiple convective rolls could also be captured using the first order UDS approximation, although, sufficiently large grid configurations is required. Dantzig [80] was the first to observe the multi-cellular feature, followed by Stella and Giangi [81] and Hannoun et al. [82]. This raised a serious question of whether the flow field inside the rectangular enclosure is ***mono- or multi-cellular?***

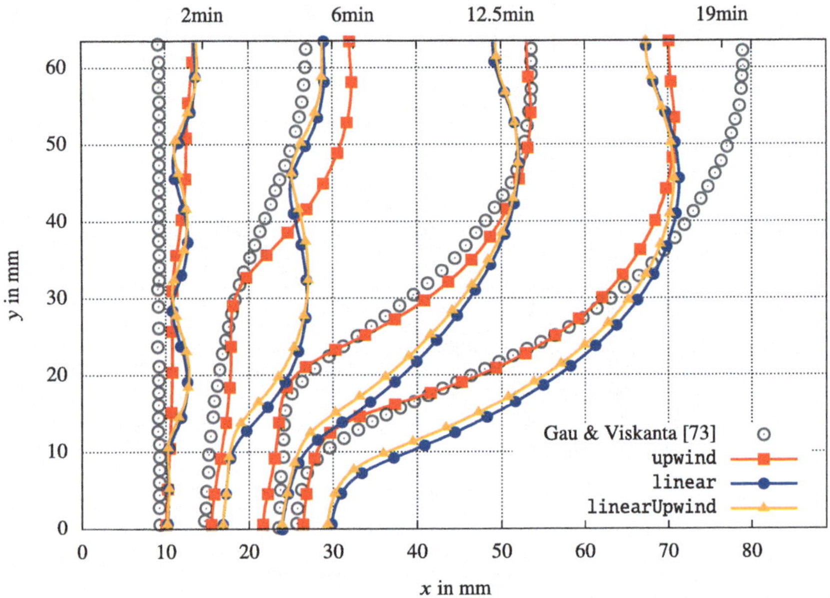

Figure 4.6 Comparison of the solid-liquid interface fronts from experiments [73] and 2D simulations for various numerical discretisation schemes at different times

In 2006 Kumar et al. [76] reported that the multi-cellular flow pattern inside the rectangular enclosure is due to the assumption of the 2D computational domain for a 3D experiment. Subsequently, they performed simulation on a 3D domain and showed that the convection in the melt is weak compared to the flow field predicted in the 2D domain, due to the presence of walls in the third dimension. Furthermore, a thermodynamically more consistent and accurate moving-grid method was employed in their investigations. More recently, Wittig and Nikrityuk [79] confirmed the existence of the secondary flow in the third direction using a fixed-grid approach.

In addition to simulations on a 2D computational domain, simulations on a 3D computational domain are also performed in this study. For the 3D computational domain, a grid with 126 × 96 × 60 cells in $x-$, $y-$ and $z-$directions (i.e., ≈ 0.725×10^6 cells) is employed. The top, bottom, front and back boundaries of the enclosure are assumed to be adiabatic. In Figure 4.7, the solid-liquid interface fronts predicted by 2D and 3D simulations are compared with the experimental findings of Gau and Viskanta [73] at different times. As observed in Figure 4.7,s the results of 3D simulations are in better agreement with the experimental data compared to the previously obtained 2D simulations. It is also noteworthy that for the same grid configuration in the $x-$ and $y-$directions, the temporal evolution of the interface front obtained from 2D simulations is wavy-like pattern during the initial stages of melting. This pattern is attributed to the multi-cellular feature, resulting in enhanced heat-transfer at the interface front [76]. Furthermore, the progression of the interface front in 2D is over-predicted compared to 3D.

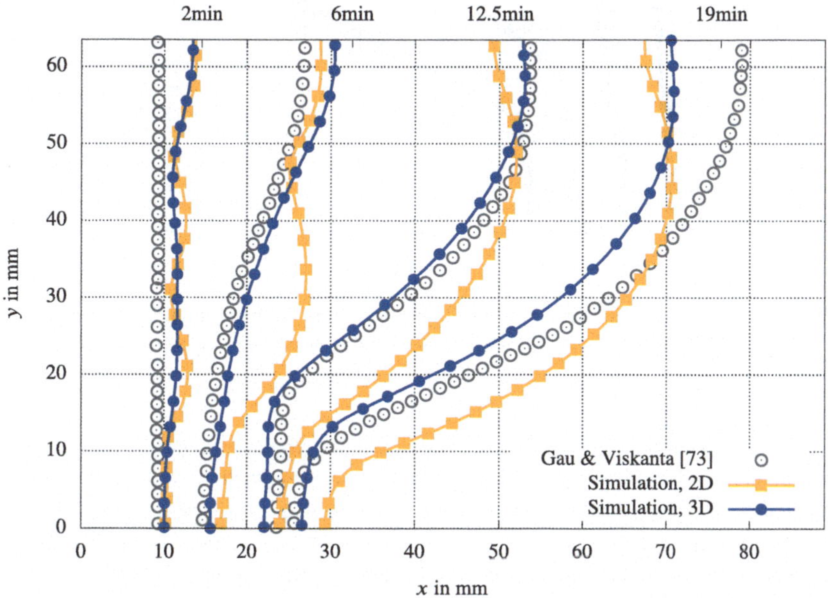

Figure 4.7 Comparison of the solid-liquid interface fronts from 2D and 3D simulations with experiments [73] at different times

Finally, the locations of the solid-liquid interface front, obtained from 3D simulations, are compared with the experimental results of Gau and Viskanta [73] and the 3D numerical results of Miehe [24], Manzke [25], Kumar et al. [76] and Wittig and Nikrityuk [79] at different times, as depicted in Figure4.8. The results of the present simulations exhibit very good agreement with the numerical results reported, particularly those based on the fixed grid approach.

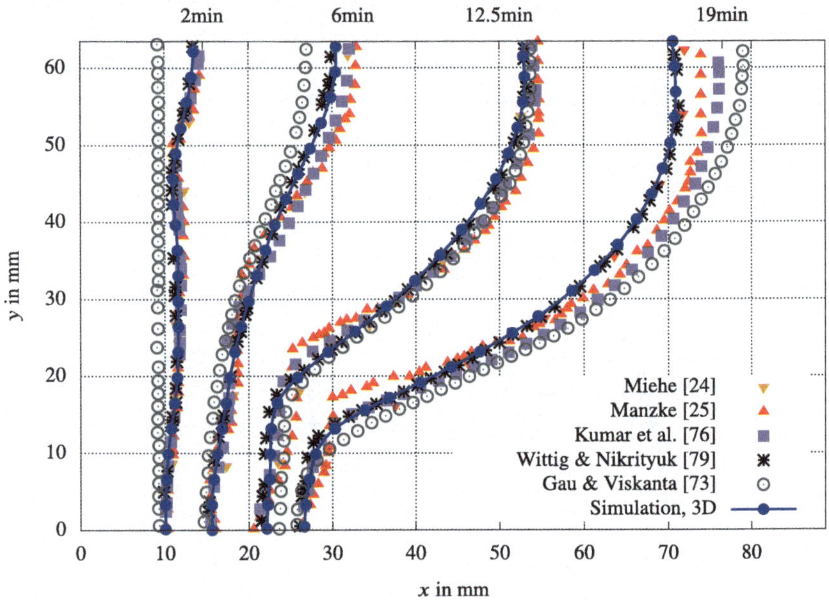

Figure 4.8 Comparison of the solid-liquid interface fronts from 3D simulation with experiments [73] and the 3D numerical results reported in the literature [24, 25, 76, 79] at different times

Further discrepancies observed between the 3D simulations results and the experiments conducted by Gau and Viskanta [73] may be attributed to the following factors: In the simulations, it is assumed that the material properties remain constant and independent of temperature. In the experiments, a pour-out method was utilised to measure the shape of the solid-liquid interface front [83]. This involved stopping the experiment at a predetermined time and rapidly draining the melted gallium from the enclosure to capture the shape of the solid-liquid interface front. However, this technique my have introduced some forced convection at the interface. Additional deviations may arise due to uncertainties in maintaining consistent wall temperatures and errors in acquiring interface data through probing or tracing methods employed in the experiments.

4.3 Free Surface Thermo-capillary Driven Flow with Phase Change

4.3.1 Problem Description

The numerical set-up of the 2D rectangular computational domain employed for the simulation is depicted in Figure 4.9. The dimensions L and H of the domain are 15 mm and 5 mm, respectively, and it is discretised into 140 $\times$ 62 hexahedral cells. The domain is filled with bismuth metal upto height H_{Bi} = 4 mm and the remaining space is filled with argon gas. The left wall is maintained at a constant temperature $T_{\mathrm{h}} = 552.55\,\mathrm{K}$ ($> T_{\mathrm{m}}$), while the right wall is set at $T_{\mathrm{c}} = 540.55\,\mathrm{K}$ ($< T_{\mathrm{m}}$), resulting in a temperature difference of $\Delta T = 12\,\mathrm{K}$ between the two walls. The top and bottom walls are initialised with a linear temperature distribution from T_{h} to T_{c}, ensuring that the isoline of the melting temperature $T_{\mathrm{m}} = 544.55\,\mathrm{K}$ is located at a distance x_{m} from the left wall. Moreover, the solidus temperature is set to $T_{\mathrm{s}} = T_{\mathrm{m}} - 0.25\,\mathrm{K}$ and the liquidus temperature is set to $T_{\mathrm{l}} = T_{\mathrm{m}} + 0.25\,\mathrm{K}$. The simulations are performed under micro gravity conditions with $g = 4.41 \times 10^{-4}\mathrm{ms}^{-2}$ and the material properties of bismuth, argon gas, and the relevant dimensionless numbers are listed in Table 4.5.

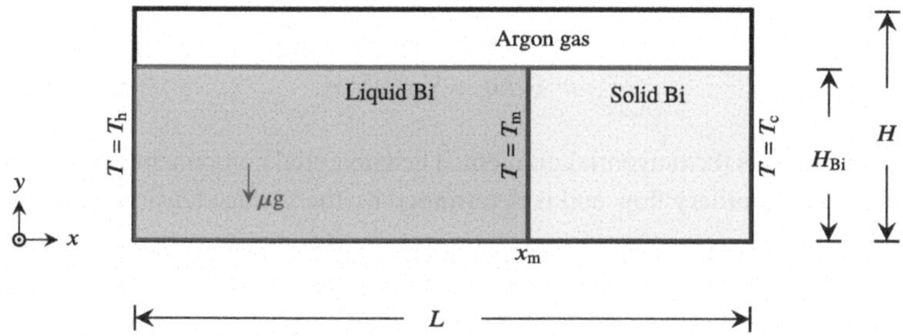

Figure 4.9 Numerical set-up of the rectangular domain filled with bismuth metal and argon gas

Table 4.5 Material properties of bismuth, argon gas and the relevant dimensionless numbers [78, 84]

Property	Symbol	Value		Unit
		Bismuth	Argon	
Density	ϱ	9780	1.6337	kg m^{-3}
Dynamic viscosity	μ	1.6×10^{-3}	2.26×10^{-5}	kg m^{-1} s^{-1}
Specific heat capacity	κ	10.35	0.0177	W m^{-1} K^{-1}
Latent heat of fusion	$\mathcal{L}$	44.6	–	kJ kg^{-1}
Thermal expansion coeff.	β	1.34×10^{-5}	–	K^{-1}
Dimensionless numbers	**Symbol**		**Value**	
Marangoni number	$Ma = (\partial\sigma/\partial t)H_{\mathrm{Bi}}\Delta T/\alpha\mu$		244	–
Capillary number	$Ca = (\partial\sigma/\partial t)\Delta T/\sigma$		2.20×10^{-3}	–
Rayleigh number	$Ra = g\beta\Delta T H_{\mathrm{Bi}}/\nu\alpha$		3.10×10^{-2}	–
Stefan number	$Ste = c_p\Delta T/\mathcal{L}$		3.30×10^{-2}	–
Bond number	$Bo = \varrho g H_{\mathrm{Bi}}^2/\sigma$		1.88×10^{-4}	–
Prandtl number	$Pr = \nu/\alpha$		1.90×10^{-2}	–

Mathematical modelling

In order to validate this case, the source term corresponding to the surface force discussed in Sect. 3.2 has to be extended. The surface tension force per unit area (f_σ) can be resolved in to two components; the normal component ($f_{\sigma,\mathrm{n}}$) and tangential component ($f_{\sigma,\mathrm{t}}$) [78, 84, 85]:

$$\begin{aligned} S_\sigma &= (f_{\sigma,\mathrm{n}} + f_{\sigma,\mathrm{t}})|\nabla\alpha|, \\ &= (\sigma k \boldsymbol{n} + \nabla_t\sigma)|\nabla\alpha|, \end{aligned} \tag{4.15}$$

were ∇_t represents the tangential gradient. The tangential component is responsible for the thermo-capillary flow and is determined by the surface tension gradient:

$$f_{\sigma,\mathrm{t}} = \nabla_t\sigma = \frac{\partial\sigma}{\partial T}[\nabla T - \boldsymbol{n}(\boldsymbol{n}\cdot\nabla T)]. \tag{4.16}$$

Combining equations (4.15) and (4.16), the surface tension source term required to predict thermo-capillary flows is obtained:

$$S_\sigma = \left\{\sigma k \boldsymbol{n} + \frac{\partial\sigma}{\partial T}[\nabla T - \boldsymbol{n}(\boldsymbol{n}\cdot\nabla T)]\right\}|\nabla\alpha|. \tag{4.17}$$

4.3.2 Results and Discussions

Figure 4.10 shows the velocity flow field of the bismuth-argon system under microgravity conditions at steady state. The linear temperature gradient across the computational domain induces thermo-capillary flow along the bismuth-argon interface (depicted by the blue line). Thus, causing the surface of the liquid bismuth metal at the left hot wall to flow towards the cold solid-liquid interface front. The liquid cools down by melting the solid and descends along the interface, rejoining the bulk liquid. This process results in a strong primary circular flow in the clockwise direction and a deformed solid-liquid interface front (illustrated by the red line). A secondary circular flow in the anticlockwise direction is also observed near the lower corner of the hot wall [84]. However, the strength of the secondary flow is relatively weak. Furthermore, the argon gas above the liquid bismuth also progresses along the free surface and thus flows in an anticlockwise direction. As illustrated in Figure 4.11, the solid- liquid interface front at steady state obtained from the simulation shows a good agreement with findings in the literature [78, 84],

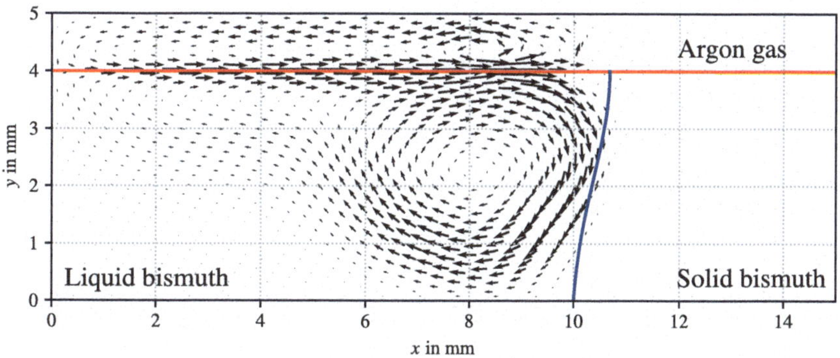

Figure 4.10 Numerical prediction of the velocity flow field, bismuth-argon interface (blue) and solid/liquid interface (red) at steady state

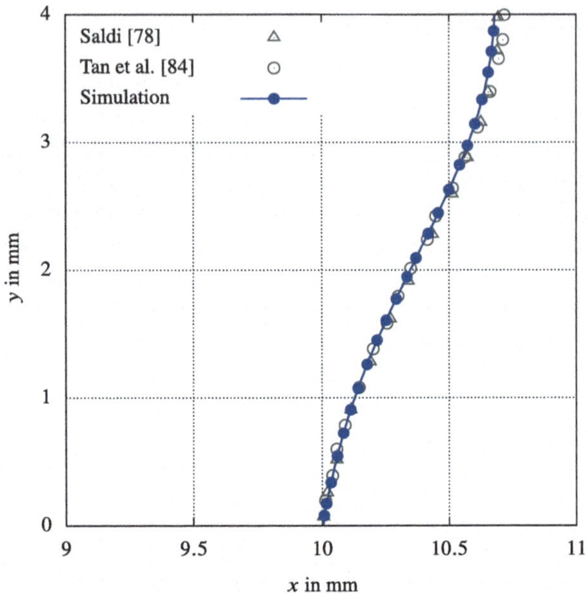

Figure 4.11 Comparison of the solid-liquid interface front at steady state with numerical results of Saldi [78] and Tan et al. [84]

moreover, the maximum deviation, when compared to the literature, is determined to be less than 1 %.

4.4 Melting of Paraffin Wax in the Presence of Gas Phase

4.4.1 Experimental Set-up

Experiments are undertaken to validate the numerical model concerning solid-liquid phase change in a multi-fluid system. Figure 4.12 illustrates the schematic of the experimental set-up adapted from [86–88]. The domain comprises a square enclosure with inside dimensions measuring 50 mm × 50 mm × 50 mm. The top, bottom, front and back walls are made of Plexiglas material, each with a thickness of 4 mm. In contrast, the left and right walls are fabricated from aluminium with a plate thickness of 6 mm and 12 mm, respectively, serving as isothermal walls. The left wall functions as a controlled heat source with a self-adhesive heating foil applied to its

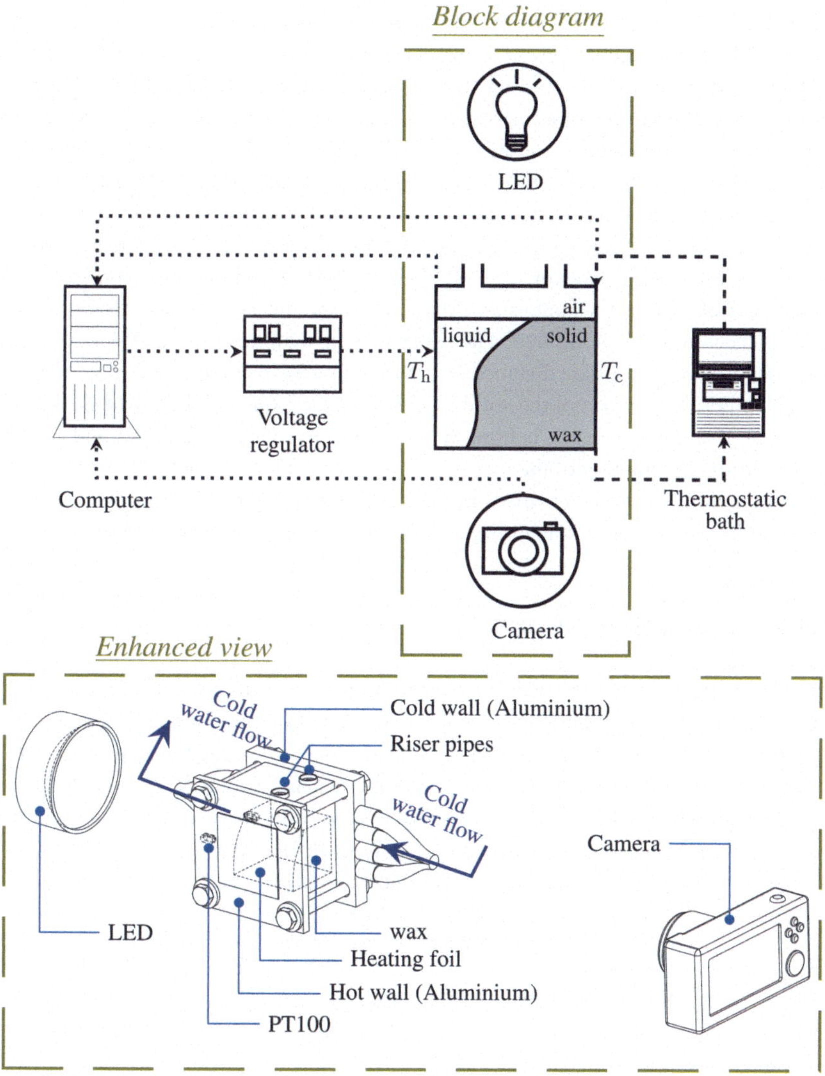

Figure 4.12 Schematic representation of the experimental set-up

outer side and is regulated through voltage and current controls. On the other hand, the opposite right wall serves as a regulated cold surface and is connected to the LAUDA Ecoline RE206 thermostat[4] for cooling. The thermostat directs cooling water through embedded channels in the metal plate, ensuring a constant temperature. Additionally, PT100 thermocouples are positioned on both the left and right walls, facilitating the transmission of the temperature data to the measurement software.[5]

To accommodate volume expansion/contraction during the phase change, multipurpose riser pipes are affixed to the top wall, allowing free airflow. The enclosure is initially filled with paraffin wax (RubiTherm® 31; RT31) in its liquid from up to a height of 40 mm through one of the riser pipes and is subsequently set aside until it has completely solidified. The experimental setup includes a DSLR camera positioned at an appropriate distance from the front wall to capture the sharp images of the temporal evolution of the solid-liquid interface front. A LED diffuser is also placed at a suitable distance behind the enclosure to identify the solid and liquid phases during the melting of the paraffin.[6] The experiment commences by elevating the temperature of the left wall from the initial temperature (T_{ini}) to the hot wall temperature (T_{h}), while maintaining the cold wall at a temperature $T_{\mathrm{ini}} = T_{\mathrm{c}}$.

4.4.2 Simulation set-up

The 2D computational domain utilised in the simulation is illustrated in Figure 4.13, and the corresponding boundary conditions are listed in Table 4.6. The left wall is maintained at a constant temperature $T_{\mathrm{h}} = 313.15\,\mathrm{K}$ ($> T_{\mathrm{m}}$) and the right wall at $T_{\mathrm{c}} = 295.15\,\mathrm{K}$ ($< T_{\mathrm{m}}$). The top and bottom boundaries are insulated, i.e., the net heat flux through these boundaries is set to zero. The solidus temperature (T_{s}) is set to $T_{\mathrm{m}} - 0.5\,\mathrm{K}$ and liquidus temperature (T_{l}) to $T_{\mathrm{m}} + 0.5\,\mathrm{K}$. At time $t = 0$, the (solid phase) wax-air system inside the enclosure has an initial temperature $T_{\mathrm{ini}} = T_{\mathrm{c}}$.

[4] The thermostat possesses an operating temperature range of −20 °C to 200 °C, with a temperature stability of ± 0.01 K [89].

[5] In this study, DASYLab® data acquisition software was employed for the purpose of regulating the temperature of the heating foil and recording the temperature data during experiments

[6] Due to the opaqueness of the the solid phase, the light can only pass through the transparent liquid phase, facilitating the easy identification of the melted portion of the wax

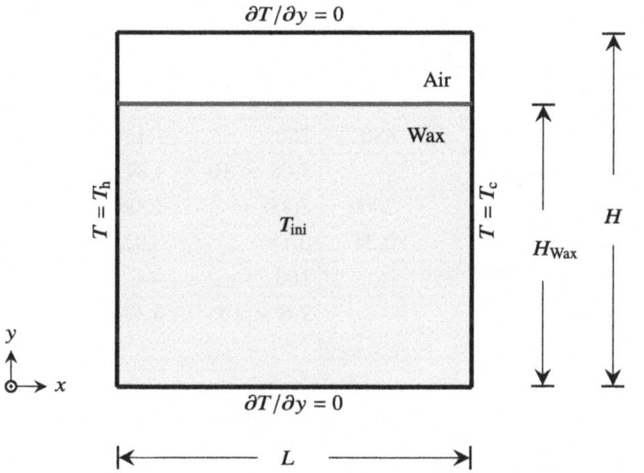

Figure 4.13 Numerical set-up for simulating the melting of wax in the presence of gas phase

Table 4.6 Boundary conditions for simulating the melting of wax in the presense of gas phase

Boundary	α	γ	T	u
	[-]	[-]	[K]	[m s^{-1}]
left	`zeroGradient`	1	313.15	0
bottom	1	`zeroGradient`	`zeroGradient`	0
right	`zeroGradient`	0	295.15	0
top	0	`zeroGradient`	`zeroGradient`	0
front	`empty`			
back	`empty`			

Material properties

The material properties of RT26 to RT35 provided by the manufacturer [90] are comparable. The specific heat capacity and the thermal conductivity of the above mentioned PCMs are the same and have uniform values for each (solid/liquid) phase. Shmueli et al. [91] compared their numerical results with experiments using the material properties provided by the manufacturer with slightly varied thermal conductivities for solid and liquid phase. The material properties used in this study are adapted from the above literature and are listed in Table 4.7.

Table 4.7 Material properties of RT31 [90, 91]

Property	Symbol	RT31		Air	Unit
		Solid	Liquid		
Density	ϱ	880	760	1.1885	$\mathrm{kg\,m^{-3}}$
Dynamic viscosity	μ	–	3.08×10^{-3}	1.81×10^{-5}	$\mathrm{kg\,m^{-1}\,s^{-1}}$
Specific heat capacity	c_p	2000	2000	1006	$\mathrm{J\,kg^{-1}\,K^{-1}}$
Thermal conductivity	κ	0.24	0.15	0.0258	$\mathrm{W\,m^{-1}\,K^{-1}}$
Latent heat of fusion	$\mathcal{L}$	–	160	–	$\mathrm{kJ\,kg^{-1}}$
Thermal expansion coeff.	β	–	7.6×10^{-4}	3.38×10^{-3}	$\mathrm{K^{-1}}$

4.4.3 Results and discussions

The grid convergence study is performed to investigate the influence of various grid configurations on the solid-liquid interface front. The solid-liquid interface fronts at time $t = 25$ min for 3 grid configurations in $x-$ and $y-$directions i.e., 175×175 cells, 275×275 cells and 375×375 cells are compared with the interface front obtained from experiments as illustrated in Figure 4.14.The height of the liquid paraffin in the experimental domain varies due to the volume expansion during phase change. Therefore, for simplicity, the melting fronts have been normalised to the width of the experimental domain in the x-direction and the height of the liquid paraffin in the y-direction.

As seen in the Figure 4.14, the deviations in the solid-liquid interface front obtained using grid configurations 275×275 cells and 375×375 cells are minuscule. Therefore, a computational grid with 275×275 cells is sufficient in terms of the computational time and resources and is chosen for further investigation.

Figure 4.15 shows the experimental and numerical, temporal evolution of the melting front at 15 min, 30 min, 45 min and 60 min. The liquid wax close to the hot wall ascends and traverses horizontally towards the cold solid-liquid interface front, where it releases heat at the upper region of the solid wax. Subsequently, the liquid paraffin descends along the cold melting front, resulting in a more rapid advancement of the melting front in the upper region compared to the lower region.

During the experiment, it is observed that the liquid wax-air interface is thicker than in simulations (as indicated by the blue dashed ellipse in the Figure 4.15). This disparity is attributed to the contact angle between the liquid wax-air interface and the Plexiglas wall, leading to the formation of a meniscus near the Plexiglas wall. In

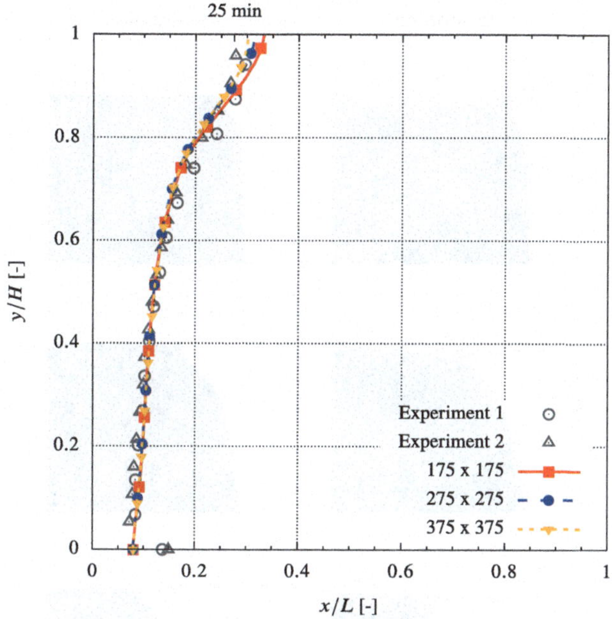

Figure 4.14 Comparison of the solid-liquid interface fronts at time $t = 25\,\text{min}$ for three distinct grid configurations

the experiment, evaluation of the solid-liquid interface front and the liquid wax-air interface is conducted through a systematic analysis of images procured from the frontal Plexiglas wall. Hence, in the experimental result, the incorporation of this meniscus contributes to the observed increase in the thickness of the liquid wax-air interface.

Figure 4.16 shows the comparison of melting fronts between experimental data and simulation. Remarkably, the numerically obtained solid-liquid interface fronts exhibit a good agreement with the experimental results. However, the discrepancy in the lower region of the enclosure ($y/H < 0.2$) observed between the interface fronts at $t = 45\,\text{min}$ and $60\,\text{min}$ is attributed to the additional heat flux from the bottom. Despite efforts to insulate the top and bottom Plexiglas walls with an additional layer of material, a minor heat flux into the bottom wall is still present, facilitated by the thermal conductivity of the Plexiglas[7], which is not considered in the simulations.

[7] The thermal conductivity of Plexiglas utilised in this study is $0.19\,\text{W}\,\text{m}^{-1}\,\text{K}^{-1}$ [92].

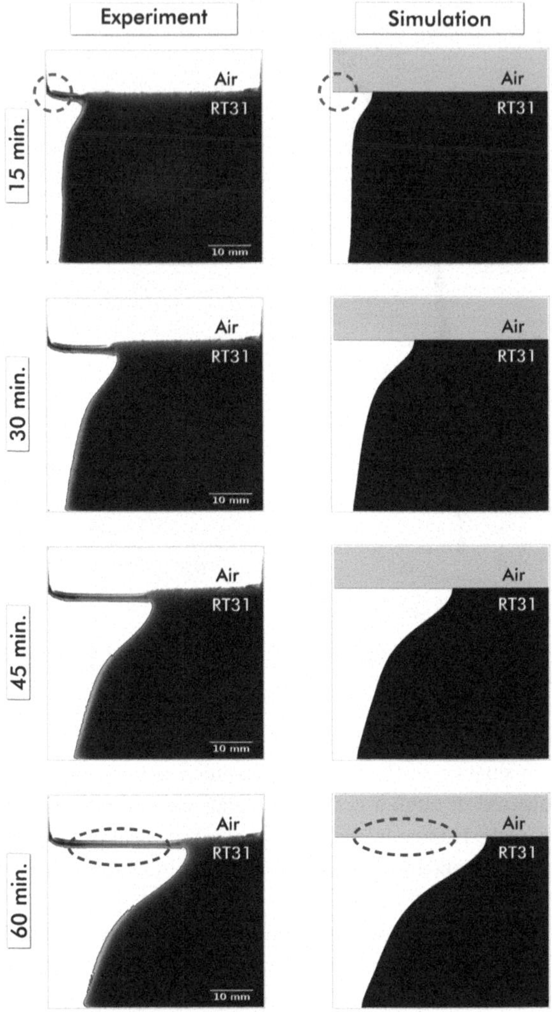

Figure 4.15 Qualitative analysis of the solid-liquid interface fronts: Experiment versus simulation

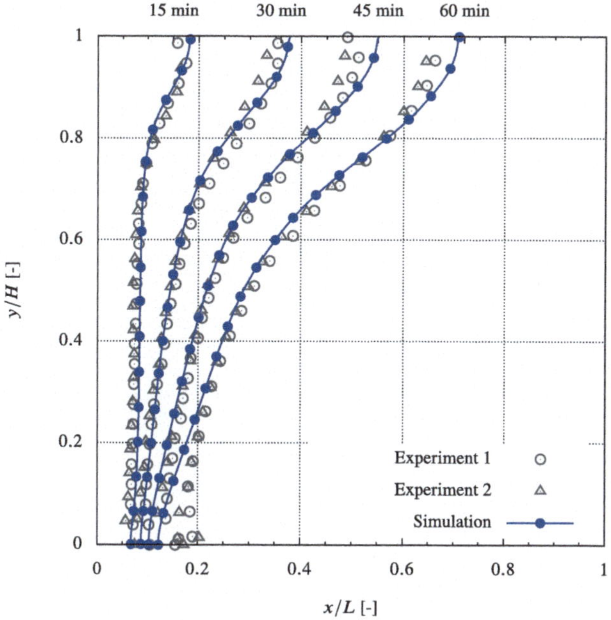

Figure 4.16 Quantitative analysis of the solid-liquid interface fronts: Experiments versus simulation

* * *

The validation of the developed numerical model was carried out using well-documented reference cases from the literature. The simulation results exhibited good agreement when compared to both analytical solution and other previously reported numerical results. Moreover, a very good agreement was also observed between the simulation results and the experimental data obtained through in-situ experiments conducted in this study. Some deviations were observed and were discussed in detail. The accuracy and reliability of the developed model were demonstrated by this validation study. The Computational Thermo-Fluid Dynamics analysis of the AT welding process can now be carried out using this model.

5 Exploring Aluminothermic Welding of Rails Through Experimental Investigations

The AT welding process brings its own set of challenges, as extremely high temperatures are involved during the process and an opaque welding setup, the visual acquisition of the physical process to understand the flow field and the solid-liquid phase change of the rails inside the casting system, is inherently difficult to observe in real-time. Thus, in this chapter, experimental trials are conducted to indirectly validate the developed numerical model discussed in Chap. 3 for predicting the flow field and solid-liquid phase changing within the mould during the pouring stage[1] of the AT welding process.

To streamline experimental parameters and avoid approximations, the welding procedure for the rapid welding process without preheating (RWOP) is employed in the present study. This facilitates a controlled investigation into the flow physics during the pouring stage, thereby providing a subtle and detailed evaluation of the accuracy of the numerical model under specific conditions. Consequently, efforts are directed towards the indirect validation of the numerical model. Experiments are conceptualised and executed to obtain the shape of the melted rails inside the mould immediately after the completion of the pouring stage, providing valuable data for validating the predictions of the numerical model.

The execution of such experiments, however, posed formidable challenges, despite the apparent straightforwardness of the adopted procedure (refer to Sect. 5.1), leading to numerous unsuccessful attempts. A total of six experiments were conducted, with the primary focus of four experiments directed towards obtaining melted rails within the mould. However, only one of these attempts proved to be successful and its findings are detailed in this study. The last two experiments were conducted to obtain the final weld profiles (i.e., the maximum width of the heat

[1] Pouring stage entails both the discharge of the molten metal from the crucible and the filling of the mould.

R. G. Kewalramani, *Computational Thermo-Fluid Dynamics of Aluminothermic Welding Process*, https://doi.org/10.1007/978-3-658-47535-2_5

affected zone (HAZ) and fusion zone (FZ)) and the temperature distribution in the rail at predetermined locations. This chapter exclusively delves into the favourable outcomes of these experiments. While the experimental results provide valuable insights, they are not adequate enough for conducting a comprehensive uncertainty analysis. This raises questions about the accuracy and reproducibility of the findings, especially concerning the shape of the melted rails.

All the experiments were performed at the Technology Innovation Center (TIC) of the Goldschmidt Holding GmbH. The experiments were skilfully conducted by Reinicke [93], with the supervision and guidance provided by Kewalramani et al. [94], ensuring the relevance of the experimental work.

5.1 Experimental Set-up and Procedure

Herein, the AT welding process is performed to obtain the shape of the melted rails inside the mould immediately after the pouring stage. A three piece mould is used instead of a traditional two piece mould to facilitate the draining of the molten metal from the mould. Moreover, a single-use reaction crucible (SRC) *a.k.a* Euro crucible, filled with 13 kg of standard Thermit-mixture (Thermit 60/Z90 E-L25) required to join the rails, is employed. Furthermore, the procedure for the RWOP process is followed.

Experiments were carried out in a sand bed with dimensions of 1250 mm × 800 mm × 250 mm, as shown in Figure 5.1. Two steel rail (type: UIC60, hardness: H260) segments, each with a length of $L = 600$ mm are aligned and clamped in the sand bed, maintaining a weld gap of 25 mm between them. The mould is mounted around the rails, covering the weld gap, and the bottom opening of the mould is sealed with a drain cover (made of thesame material as the mould) balanced on a drain plug. Such that, the drain plug and cover can be pulled (indicated by the yellow arrow in the figure) as needed. Subsequently, the SRC is positioned just above the mould and the Thermit-mixture is ignited in the crucible. A highly exothermic AT reaction occurs, producing approximately 7 kg of liquid Thermit-steel and 6 kg of alumina (slag) in the process. The molten metal (steel-slag) is discharged from the crucible and poured into the mould.

The shape of the melted rails inside the mould after the completion of the pouring stage followed by the draining process[2] is shown in Figure 5.2. As observed in the figure, the rails inside the mould undergo a melting process, resulting in nearly

[2] The drain plug and drain cover were pulled as soon as the molten metal reached at the top of the mould during the filling process.

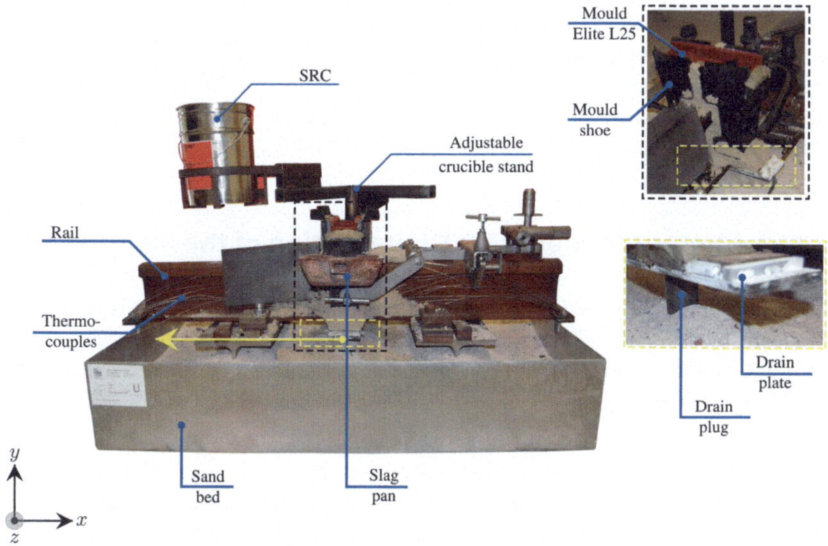

Figure 5.1 Experimental set-up to obtain the shape of the melted rails post-pouring stage

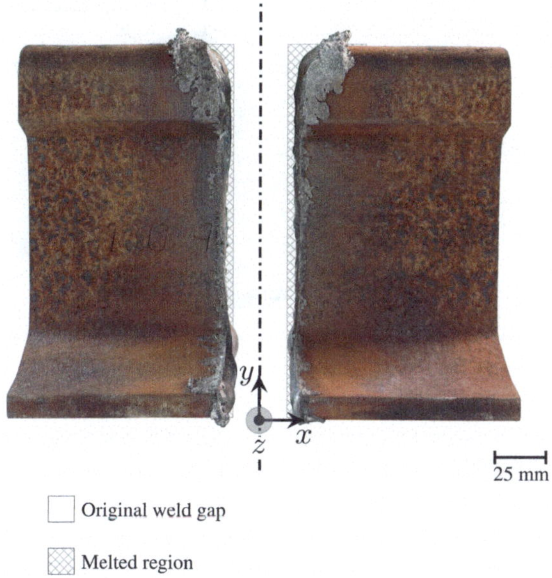

Figure 5.2 The shape of the melted rails obtained immediately after the pouring stage

symmetrical rail profiles. The melting phenomena can be attributed to two primary factors: firstly, the temperature of the molten metal produced ($T_P \approx 2473\,K$) following the Thermit-reaction surpasses the melting point ($T_m \approx 1695\,K$) of the rails and secondly, the forced convection experienced during the mould-filling stage place a crucial role in enhancing the heat transfer between the molten metal and the rails.

5.2 Photogrammetric 3D Digitalisation of the Melted Rails

In this study, the digital models of the experimentally obtained melted rails are generated using the commercially available photogrammetry software Agisoft Metashape [95]. A turntable set-up is designed for scanning small and medium sized objects for photogrammetry reconstruction, as depicted in Figure 5.3[3].

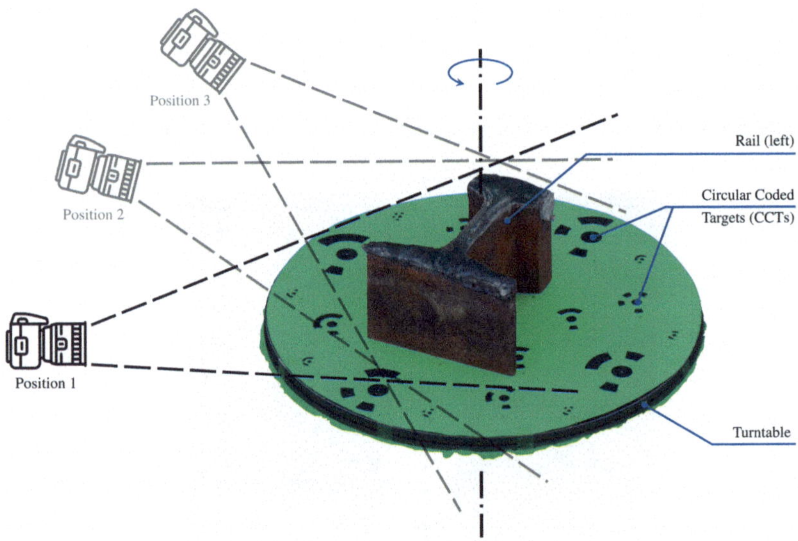

Figure 5.3 A turntable set-up for capturing a series of overlapping photographs for the three dimensional reconstruction of the rail (Experiment: left rail)

A green sheet with strategically placed circular coded targets (CCTs) is affixed to the turntable, serving as reference points for the coordinate system and scale

[3] The Camera image was acquired from SVGRepo [96]

definition of the object within the software to ensure good quality and accurate reconstruction.

The camera is placed at an appropriate distance from the centre of the turntable, and the camera angle and position can be adjusted to achieve better access to the object. The autofocus function of the camera is disabled. The key to acquiring good photos is the lighting set-up to prevent casting strong shadows on the table or background as the object is rotated, which could influence the digital reconstruction process. To achieve this, four diffused lighting sources are employed—two positioned on either side of the object, and the remaining two projected from the front. Additionally, a green screen is positioned behind the object to create a plain background.

Both rails (left and right) are individually placed on the turntable, and a series of overlapping photographs[4] are captured from three different positions/angles by manually rotating the table as evenly as possible. Approximately 150 high-quality photographs are taken for each rail, ensuring the comprehensive coverage of the complex and intricate shape of the melted region.

The workflow for image processing and 3D reconstruction in Agisoft Metashape to obtain the digital model of the rail is depicted in Figure 5.4. The photographs are initially imported in software and aligned based on accuracy parameters (lower, low, medium and higher), yielding in a sparse point cloud, along with the assumed positions of the photographs as shown in Figure 5.5(a). Subsequently, the CCTs are automatically (as the CCTs in Figure 5.3, are generated within the software)

Figure 5.4 Workflow for image processing and 3D reconstruction in Agisoft Metashape [95]

[4] High-quality photographs are captured using Sony Alpha 6000 camera

or manually (if required) identified on all the photographs, with their coordinates imported. This process results in calibrated camera positions to retrieve scaled and optimised point cloud data.

Following this, a dense cloud is generated based on quality parameters (low, medium and high) as shown in Figure 5.5(b). The coloured visualisation of the quality of the dense point cloud (referred to as dense cloud confidence), which indicates the total number of images utilised to acquire accurate details of a particular region in the geometry, is illustrated in Figure 5.5(c). Blue and green colours represent more accurate points, indicating the use of the maximum number of images to generate a point. Conversely, less accurate points are represented by red. In general, the points marked with red can be filtered out to obtain an accurate model. Subsequently, a polygon mesh is created based on the dense point cloud as shown in Figure 5.5(d). Finally, Figure 5.5(e) represents the texture-mapped digital model of the rail.

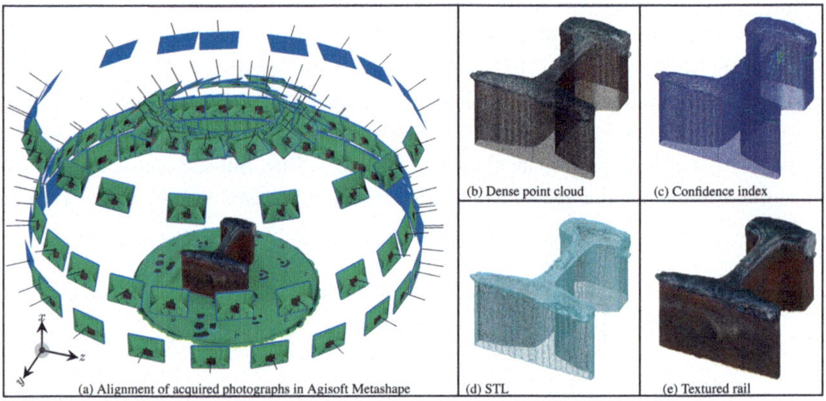

Figure 5.5 Photogrammetry of the experimentally obtained melted rails within the mould—**Note:** Herein, the photographs are acquired from four different positions

5.3 Shape of the Melted Rails

The qualitative analysis of the 3D digitalised melted rails post-pouring stage, followed by the draining of the molten metal, is shown in Figure 5.6, alongside the experiment. The digitalised representation of the melted rails shows notable consistency with the experiment. The region within the blue dashed box in Figure 5.6(c) and Figure 5.6(d) highlights the intensified melting of rails during the filling process, particularly at the railhead-web intersection where the molten metal first contacts the rail. Notably, the melting at the centre of the rail base is more dominant, which indicates the influence of the mould draining process. It is noteworthy that further melting of the rails may occur due to forced molten metal convection during the draining process.

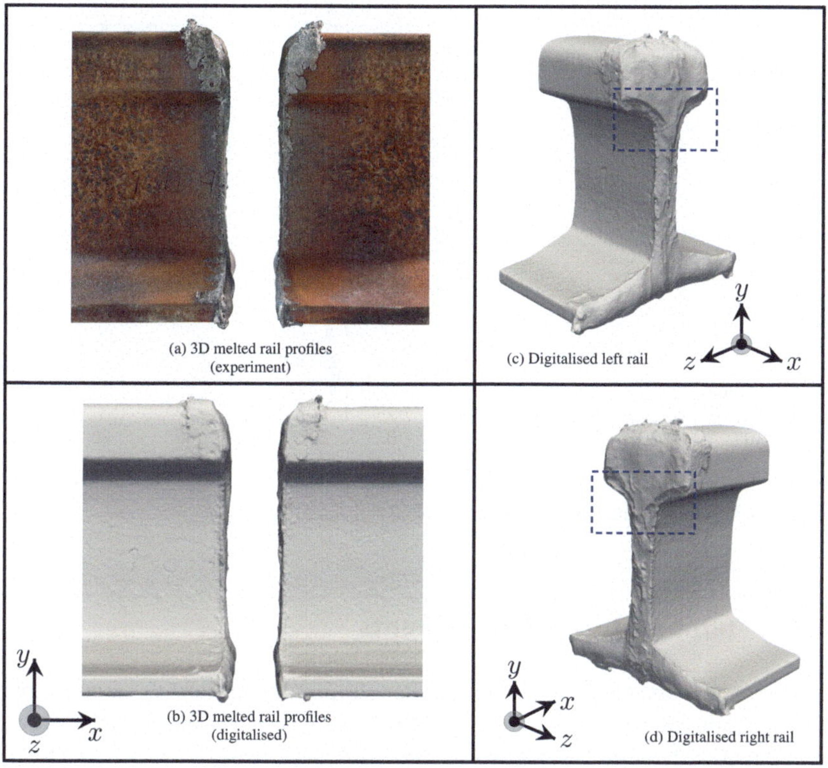

Figure 5.6 Qualitative analysis of the shape of the melted rails within the mould post-pouring stage: Experiment and Photogrammetry

The digitalised models are instrumental in determining the width of the gap obtained after the conclusion of the draining process. Figure 5.7 shows the evaluation of the width of the gap on three different longitudinal planes, i.e., at $z = -7.5\,\mathrm{mm}$,

$z = 0$ mm and $z = 7.5$ mm. The two vertical dashed black lines in the figure represents the original rail. Finally, this data will be used to validate the numerical results obtained form the simulations of the thermal pouring stage.

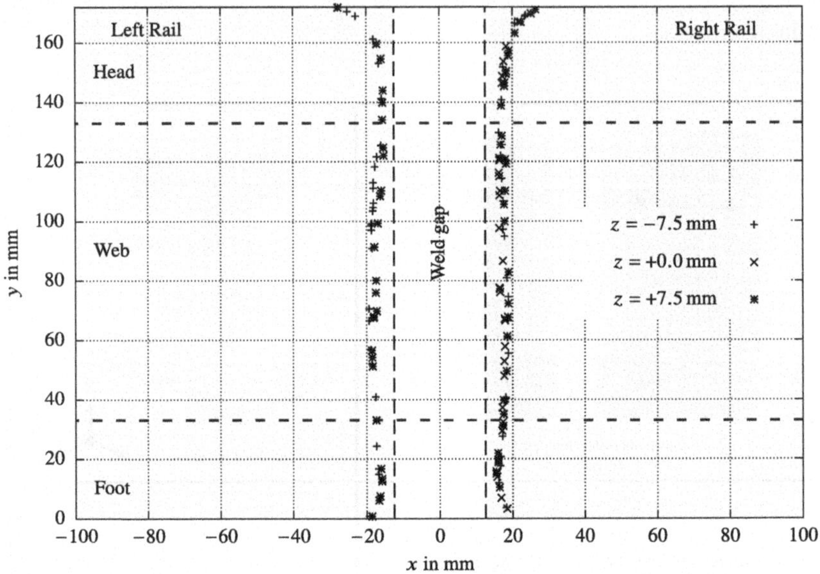

Figure 5.7 Width of the gap between the melted rails post-pouring stage

5.4 Temperature Measurements

As discussed in the beginning of this chapter, the primary focus of the experiments was to obtain the melted rail profiles within the mould post-pouring stage. However, the brief duration of each experiment, lasting less than 15 s, was found to be insufficient for obtaining comprehensive thermal distribution in the rails. To address this limitation, additional experiments were conducted specifically to measure the temperature distribution in the rail at predetermined locations.

For this purpose, holes of 1.6 mm diameter holes were drilled from the outer surface of both rails at three locations along the horizontal direction i.e., from the rail end within the mould, and six locations in the vertical direction, namely, head-centre (H1), head-outer (H2), web-centre (W1), foot-centre (F1), foot-outer-top (F2) and foot-outer-side (F3), as shown in Figure 5.8. The coordinates for these measurement locations from the weld centre are listed in Table 5.1. Each rail was equipped with 18 K-type (NiCr-Ni) mantle thermocouples, each having a diameter of 1.5 mm, to observe both temporal and spatial evolution of the temperature field within the rail. The thermocouples were securely held in position with luting paste. To facilitate the acquisition of data, temperature readings were recorded at a frequency of 1 Hz using a 56-channel thermocouple input module.

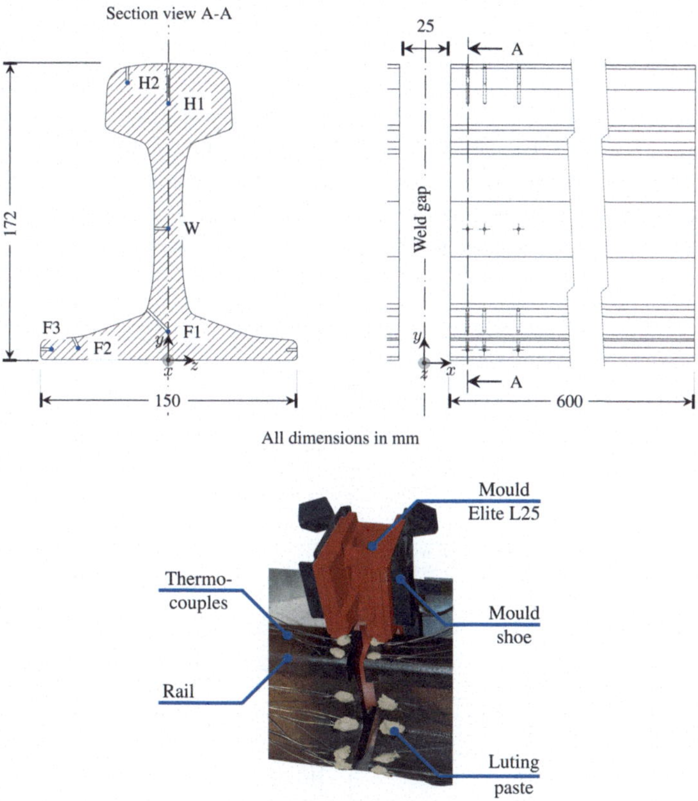

Figure 5.8 Measurement locations of thermocouples: schematic representation (top, adapted from [9]), experiment (bottom)

Table 5.1 Measurement locations of thermocouples

Location	$\pm x$ (both rails)	y	z
	[mm]	[mm]	[mm]
H1	22.5, 32.5, 52.5	149.5	0
H2		161	−25
W1		76.3	0
F1		17	0
F2		5.8	−52.5
F3		5.8	−69

Figures 5.9 and 5.10 depict the temporal evolution of the temperature in the rail head, web and base regions, respectively, during the cooling/solidification stage. It is noteworthy that the thermal information at certain locations in the rail was not recorded due to thermocouples being detached or broken either during the of

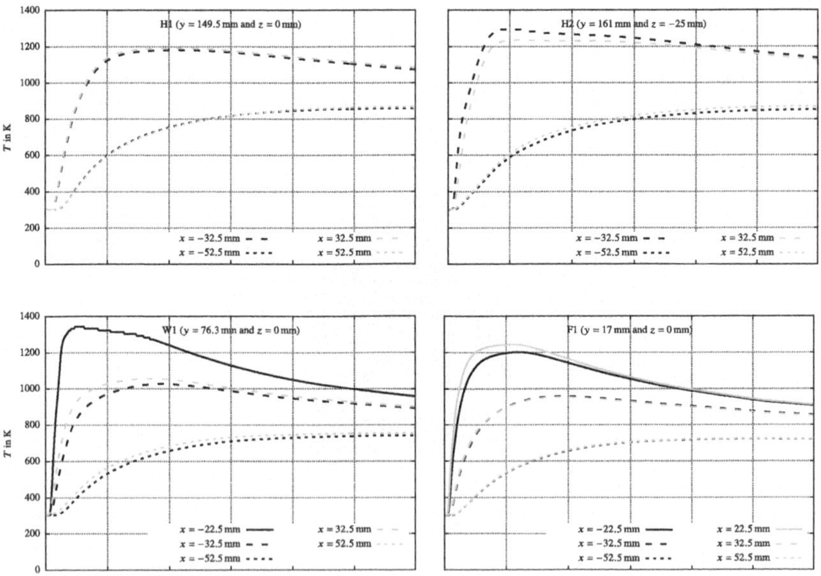

Figure 5.9 Temperature distribution in the rail cross head (H1, H2), web (W1) and base (F1) regions during mould filling and cooling stages

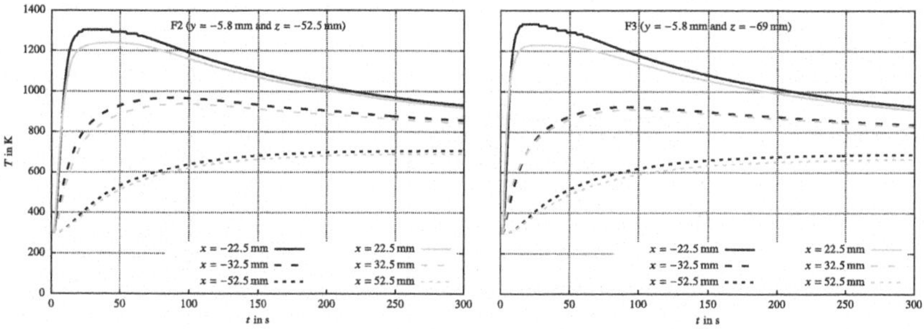

Figure 5.10 Temperature distribution in the rail at base (F2, F3) regions during mould filling and cooling stages

mounting the mould or when sealing the casting set-up with sand to prevent molten metal from leaking. Additionally, some thermocouples were also damaged upon contact with the molten metal.

5.5 Fusion and Heat Affected Zone

As outlined in Sect. 5.1, two rail segments, each spanning 600 mm, were employed for the welding process, resulting in a cumulative rail length of approximately 1225 mm[5] post-welding. Subsequently the rail joint segment was cut off from the parent rail, ensuring that the weld region remained centrally positioned, as shown in Figure 5.11.

The rail segment was sectioned longitudinally to analyse the metallurgical characteristics, encompassing both macro- and micro-structure[6] of the welded region. The longitudinally sectioned half of the rail segment was subjected to metallurgical surface finish processes, i.e., polishing and etching. Figure 5.11(a) provides a visual representation of the metallographic analysis of a typical rail joint, revealing three different zones, namely, the fusion zone (FZ), the heat-affected zone (HAZ) and unaffected base metal (BM). The inner white curves represent the maximum width of the FZ, while the outer white curves represent the extents of the HAZ. Both the FZ and HAZ exhibit variations along the rail height and the dimensions of the maximum width of the FZ and HAZ at different locations were deduced based on the schematic representation in Figure 5.11(b). The experimentally obtained maxi-

[5] Including the weld gap of 25 mm.

[6] **Note:** The micro-structures were not examined in the scope of this thesis.

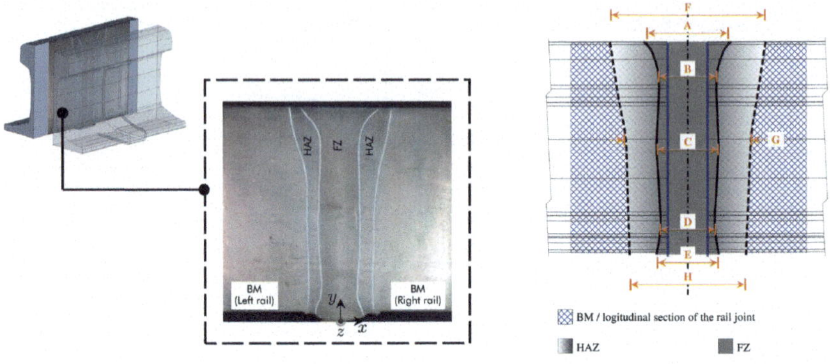

(a) Experiment (b) Schematic view (adapted from [9])

Figure 5.11 Metallographic analysis of the longitudinal section across the rail joint [97]

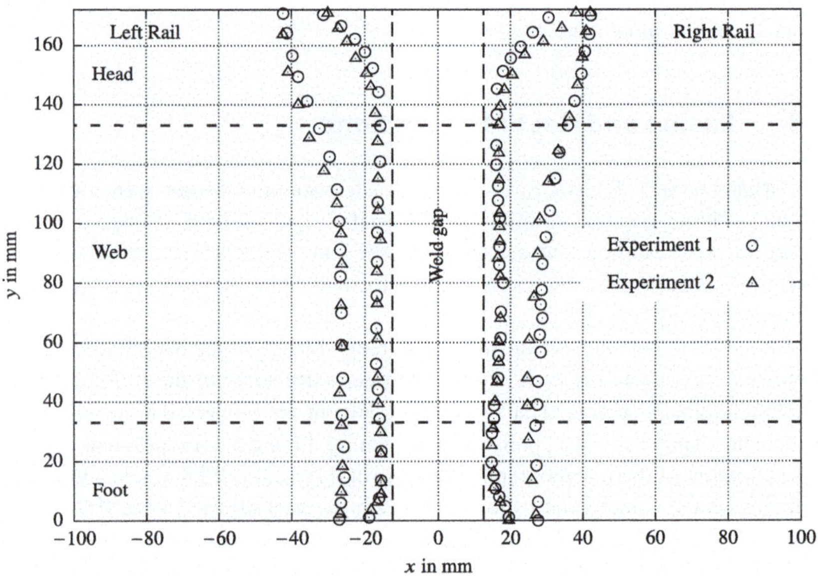

Figure 5.12 Experimentally obtained maximum width of FZ and HAZ

mum width of the FZ and HAZ from two experiments are digitalised using the tool WebPlotDigitizer [98] and is illustrated in Figure 5.12, while their respective weld measurements, as depicted in Figure 5.11(b), are summarised in Table 5.2.

Table 5.2 Measurements of the FZ (A-E) and HAZ (F-H)

		Experiment 1	Experiment 2
		W in mm	W in mm
FZ	A	65	69
	B	32	34
	C	34	33
	D	28	32
	E	39	39
HAZ	F	84	85
	G	55	53
	H	55	55

* * *

The results presented in this chapter, encompassing the shape of the melted rails post-pouring stage, the temporal and spatial evolution of the temperature field in the rails at predetermined locations during the entire welding process, and the final weld profiles, specifically detailing the shape and dimensions of the maximum width of the FZ and HAZ are utilised as validation purposes in the numerical predictions outlined in Chaps. 6 and 7.

Simulation of the Thermal Pouring Stage

6

This chapter focuses on the numerical exploration of the thermal pouring stage of the AT welding of rails, which includes crucible discharge and mould filling. The solid-liquid phase change phenomena and RANS $k - \varepsilon$ equations are solved, as detailed in Chap. 3. The shape of the melted rails obtained from the numerical simulations are compared to those from experiments. Moreover, the influence of pouring temperature on the shape of the melted rails within the mould post-pouring stage is studied. In the experiments described in Chap. 5, the melted rails within the mould post-pouring stage are obtained by employing a draining mechanism. Consequently, the influence of the draining of the molten metal on the shape of the melted rails is also examined. It is noteworthy that some of the results presented in this chapter have been published in Kewalramani et al. [99] and the findings are built upon and extended in the current study, offering additional insights.

6.1 Simulation Set-up

Mesh generation

In this study, Hexahedra-dominant computational grids are generated through the use of the utilities `blockMesh` and `snappyHexMesh`, which are provided with OpenFOAM. A background mesh with hexahedral cells is initially created using the utility `blockMesh` enclosing the entire computational domain (Figure 6.1a,i). Subsequently, the existing mesh is refined by splitting the hex cells based on the refinement level in specific regions. The cells outside the computational domain are then deleted (Figure 6.1a,ii) in a process known as *castellation*. Following castellation, the mesh is smoothed and adapted to the surface of the computational domain,

R. G. Kewalramani, *Computational Thermo-Fluid Dynamics of Aluminothermic Welding Process*, https://doi.org/10.1007/978-3-658-47535-2_6

this process is referred to as *snapping* (Figure 6.1a,iii). If necessary, boundary layers can be created. Mesh refinements are strategically applied to refine the mesh locally and wherever necessary, aiming to reduce the total number of cells in the computational domain.

The mesh strategy employed in this work is illustrated in Figure 6.1b. The region of interest is depicted in orange, with a relatively fine cell size[1] assigned compared to the red regions, the orange region is the region of interest. By adopting this strategy, the total number of cells in the domain is reduced by a factor of five, resulting in the efficient utilisation of available computational resources to address the complexities of the problem at hand.

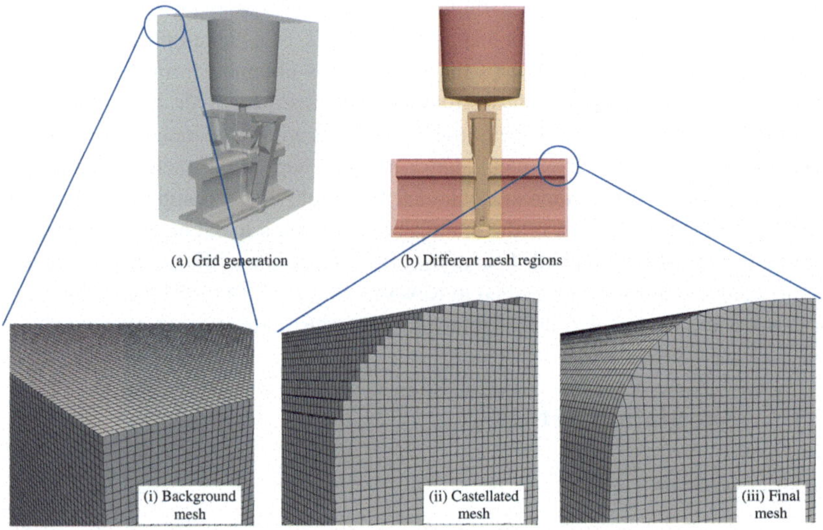

Figure 6.1 Mesh generation strategy implemented for simulating pouring process

Initial conditions and boundary conditions

Figure 6.2 illustrates the initial and boundary conditions applied in the simulations. The crucible is initialised with the liquid Thermit-steel (depicted in orange) at a temperature of 2473 K. The rails (depicted in grey) are initialised with a melt fraction

[1] Here, $\Delta x = \Delta y = \Delta z$, unless specifically mentioned otherwise

$\gamma = 0$, maintaining a distance of approximately 25 mm. Additionally, the rails are assumed to be at an ambient temperature of 293 K. At time $t = 0$ s, the entire system is at rest i.e., $\boldsymbol{u} = 0\,\mathrm{m\,s^{-1}}$. The domain boundaries are defined by the inlet (depicted in blue), outlet (depicted in green), sand plug (depicted in black) and walls (depicted in transparent grey).

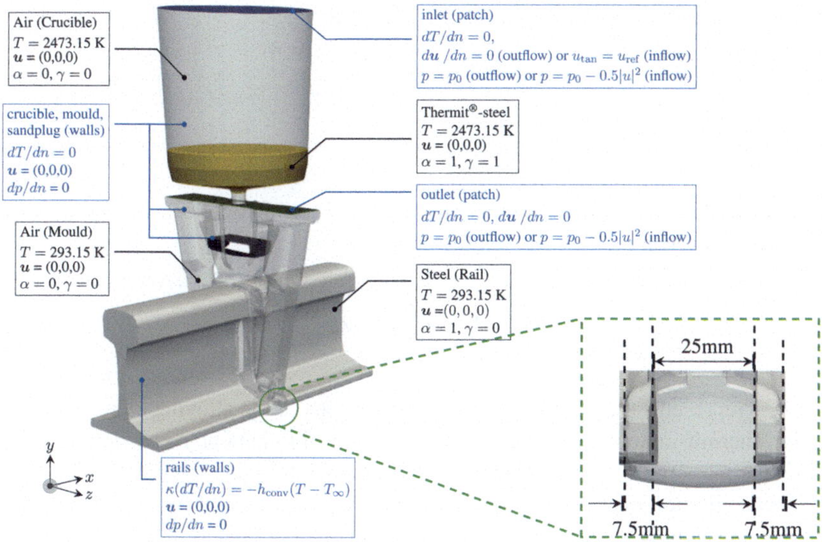

Figure 6.2 Initialisation of field parameters, domain boundaries and boundary conditions [99]

Material properties

The relevant material properties of the steel, including Thermit-steel in its liquid and and rail steel in its solid state employed in the CFD Simulations, are listed in Table 6.1. Given that the study focuses on investigating the RWOP process, the thermophysical properties of air at temperature $T_{air} = 293.15$ K [100] are utilised in the simulations.

Table 6.1 Material properties of the steel [9, 13, 101]

Property	Symbol	Value	Unit
Density	ϱ	6900	kg m^{-3}
Dynamic viscosity	μ	5×10^{-3}	kg m^{-1} s^{-1}
Specific heat capacity, solid	$c_{p,\text{sol}}$	640	J kg^{-1} K^{-1}
Specific heat capacity, liquid	$c_{p,\text{liq}}$	680	J kg^{-1} K^{-1}
Thermal conductivity, solid	κ_{sol}	30	W m^{-1} K^{-1}
Thermal conductivity, liquid	κ_{liq}	55	W m^{-1} K^{-1}
Latent heat of fusion	$\mathcal{L}$	277×10^3	J kg^{-1}
Surface tension	σ	1.5795	N m^{-1}

The solidus and liquidus temperature of the Thermit-steel can be determined using the empirical relationship expressed as:

$$T = \bar{T} - \left[\sum_j a_j \chi_j \right], \tag{6.1}$$

where j represents the range of mass concentrations of the alloying elements used in the Thermit-mixture. The regression coefficient for element j is denoted as a_j, while the mass fraction of element j in % is denoted by χ_j.

The regression coefficient of the alloying elements, crucial for calculating the solidus and liquids temperatures of the Thermit-steel are listed in Table 6.2 [8, 9, 102]:

Table 6.2 Regression coefficients of the alloying elements; published in [97]

	Alloying elements (j)										
	$\bar{T}$	C	Si	Mn	P	S	Ni	Cr	Al	Cu	Ti
	°C										
T_{sol}	1535	200	12.3	6.8	124	183	4.3	1.4	4	0	0
T_{liq}	1537	88	8	5	30	25	4	1.5	0	5	18

Thus, the solidus temperature and liquidus temperature for the rail material R260 are estimated to be $T_{sol} = 1380\,°C$ ($\approx 1653\,K$) and $T_{liq} = 1465\,°C$ ($\approx 1738\,K$), respectively [9].

Finally, the contact angle values employed to describe the wetting phenomena of liquid Thermit-steel within the welding (casting) system, namely the crucible and mould (composed of waterglass-bonded SiO_2) and sand plug (composed of waterglass-bonded Al_2O_3-SiO_2) are given in Table 6.3 [16, 18]:

Table 6.3 Contact angle θ values of liquid Thermit-steel with crucible, mould, sand plug and rails used for the simulations

Boundary	Value
crucible	131°
mould	131°
sand plug	134°
rails	90°

Numerical parameters

The numerical schemes employed for the simulations are summarised in Table 6.4. Additionally, the time step is determined by maintaining a maximum Courant number (Co) below 0.2, leading to a typical time step of the order $\Delta t = 10^{-6}\,s$. Finally, the convergence criteria are met when the normalised residual values of the continuity, momentum and energy equations are below 10^{-8}, 10^{-10} and 10^{-12}, respectively.

Table 6.4 Numerical discretisation schemes

Term	Operator	Scheme	OpenFOAM
Temporal	$\frac{\partial \varphi}{\partial t}$	Euler	`Euler`
Gradient	$\nabla \varphi$	Central differencing	`Gauss linear`
Divergence	$\nabla \cdot (\varrho \overline{\boldsymbol{u}} \overline{\boldsymbol{u}})$	Self-filtered central differencing	`Gauss SFCDV`
	$\nabla \cdot (\varrho c_p \overline{\boldsymbol{u}} \overline{T})$	Limited linear differencing	`Gauss limitedLinear 1`
	$\nabla \cdot (\gamma \overline{\boldsymbol{u}})$	van Leer limiter	`Gauss vanLeer01`
	$\nabla \cdot (\alpha \overline{\boldsymbol{u}})$	van Leer limiter	`Gauss vanLeer01`
	$\nabla \cdot (\alpha \boldsymbol{u}_r)$	Interface compression	`Gauss interfaceCompression`
	$\nabla \cdot (\overline{\boldsymbol{u}} k)$	Upwind differencing	`Gauss upwind`
	$\nabla \cdot (\overline{\boldsymbol{u}} \varepsilon)$	Upwind differencing	`Gauss upwind`
Laplacian	$\nabla \cdot (\nabla \varphi)$	Central differencing Explicit non-ortho. correction	`Gauss linear corrected`

6.2 Results and Discussions

6.2.1 Mesh Sensitivity Analysis

To investigate the influence of the computational mesh on the flow field within the mould, simulations are carried out using five distinct cell sizes i.e., Δx = 1.50mm, 1.25mm, 1.00mm, 0.75mm and 0.50mm, respectively, within the specified region of interest (as depicted by orange region in Figure 6.1b). The total size of the computational mesh for each cell size is detailed in Table 6.5.

Table 6.5 Total size of the computational domain for each Δx investigated in this study

Cell size	cells
mm	$\times 10^6$
1.50	1.24
1.25	2.07
1.00	3.70
0.75	7.78
0.50	23.67

However, due to the unavailability of appropriate information from experiments that could be utilised to assess grid independence, the positions of the liquid Thermit-steel inside the mould for various grids at $t = 0.35$ s are examined, as illustrated in Figure 6.3.

Deviations in the positions of the liquid Thermit-steel inside the mould between the mesh of cell size $\Delta x = 0.75$ mm and the mesh of cell size $\Delta x = 0.5$ mm are found to be minuscule. Consequently, a computational mesh comprising 7.78 million cells with $\Delta x = 0.75$ mm is chosen for further investigations.[2]

[2] **Note:** For the mesh independent study, the crucible was filled with 7 kg of liquid Thermit-steel.

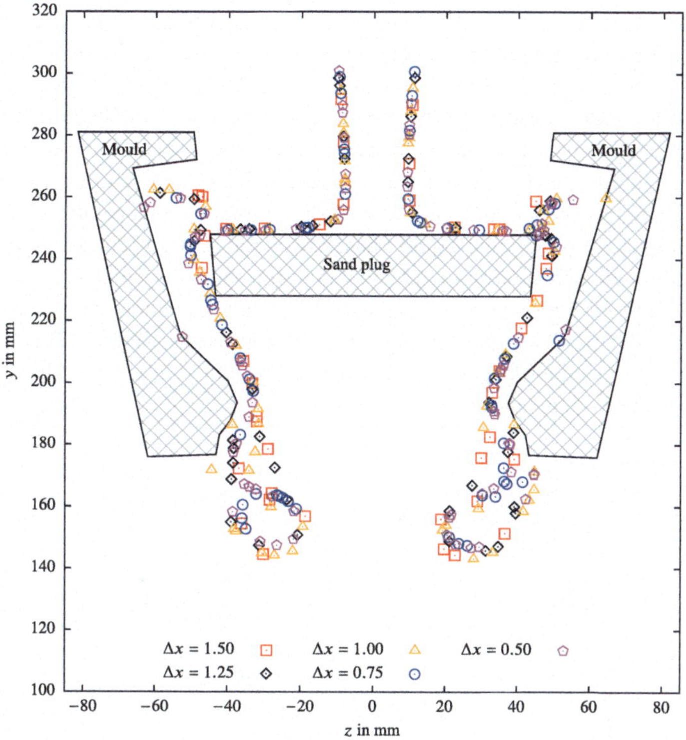

Figure 6.3 Comparison of the flow filed inside the mould at time $t = 0.35$ s for different values of Δx in mm

6.2.2 Computational Speed-up

The HPC-cluster at TU Bergakademie Freiberg, boasting a peak performance of 307.2 TFlops, is utilised for the numerical simulations reported in this study. The computational mesh, comprising 7.78 million cells, is subdivided into subdomains using the Scotch [103] domain decomposition approach. Each subdomain is then assigned to a CPU-core (or processor). Figure 6.4 demonstrates the speed-up performance of the developed multi-fluid, multiphase numerical model for simulating the thermal pouring stage across various core allocations – 1, 2, 4, 8, 16, 32, 64, 128, 256 and 512 cores. A linear speed-up, resembling the ideal speed-up line, is observed up to 8 cores. However, a dip in the speed-up occurs from 8 cores onward until about 64 cores. Subsequently, the speed-up proceeds in parallel to the linear ideal line, maintaining an offset of approximately half the speed-up of the ideal line. The dip and decline in speed-up observed from 8 cores onwards can be attributed to several factors. Firstly, the allocated memory available on each core might be insufficient, especially in the transition from 8 cores to 64 cores. Secondly, an increase in the number of cores results in elevated communication times between them, leading in a reduction in speed-up. Lastly, limitations on parallelisation may also be imposed by the solid-liquid phase change algorithm (and the calculation of the additional source terms) implemented in this study. Despite achieving a notable speed-up of approximately 252 with 512 cores, the simulations are carried out using 256 cores, resulting in a speed-up of approximately 143. This decision aimed to circumvent prolonged waiting times on the HPC-cluster, ensuring efficient computational resource utilisation.

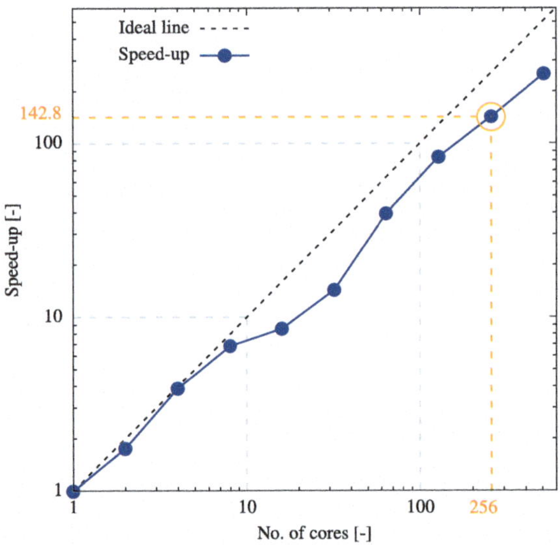

Figure 6.4 Computational speed-up obtained for the multiphase mould filling simulation of AT welding process on the HPC-Cluster of the TU Bergakademie Freiberg

6.2.3 Influence of the Pouring Temperature

The measurement of the temperature of molten metal produced after the Thermit-reaction proves to be extremely challenging, and readings obtained using a thermal camera undergo significant fluctuations [9, 22]. Consequently, four different pouring temperatures inside the crucible—$T_P = 2073$ K, 2273 K, 2473 K and 2673 K are systematically studied to understand their impact on the mould-filling process and the final melted rail profiles within the mould post-pouring stage.

Figure 6.5 shows the velocity vectors in the liquid Thermit-steel and the temperature field in the rail, along with the solid-liquid phase change during the mould-filling process at a pouring temperature of 2073 K. The stream of liquid steel, discharged from the reaction crucible initially hits the sand plug within the mould and is subsequently split into two streams. The rails are melted by the liquid streams as they flow down further into the mould along the mould walls, filling up the gap between the rails. Additionally, due to the intricate flow path within the mould, air bubbles are introduced during the mould-filling process. Almost all of the air bubbles are capable of escaping from the liquid Thermit-steel by the end of the pouring stage. However, a large air bubble is trapped just above the railhead region (refer to $t = 3$ s in Figure 6.6), where the liquid fill height exceeded the railhead height. It continues to struggle to escape from that point onwards. Finally, the melting off of the parent rail inside the mould during the pouring stage is also depicted.

Figure 6.6 illustrates the temperature distribution in the liquid Thermit-steel within the mould during the pouring stage. It is noteworthy that the casting mould and sand plug are not physically considered in the simulations, instead, they are treated as adiabatic boundary conditions. Consequently, heat is lost by the liquid Thermit-steel upon its initial contact with the rail at the railhead-web region, specifically at the head-side-riser where the liquid streams combine to fill the weld gap. As the liquid Thermit-steel flows along the rail web, it melts the rail in its path and experiences further heat loss. As a result, the bottom of the mould is filled with the liquid Thermit-steel at a temperature lower than the pouring temperature.

Similarly, Figures 6.7, 6.9 and 6.11 illustrate the flow field and the melting off of the parent rail within the mould for the remaining pouring temperatures. It is observed that there is no significant impact on the mould filling process up to $t = 0.40$ s. This lack of influence is attributed to the assumption that the thermophysical properties considered in this study remain constant for temperatures higher than the liquidus temperature of the liquid Thermit-steel. However, beyond this timeframe, a dependence of the flow field on the amount of material melted from the parent rail is observed. Nevertheless, the entrapment of the air bubble just above the railhead region is a recurring issue in all cases. Additionally, the maximum velocity of the

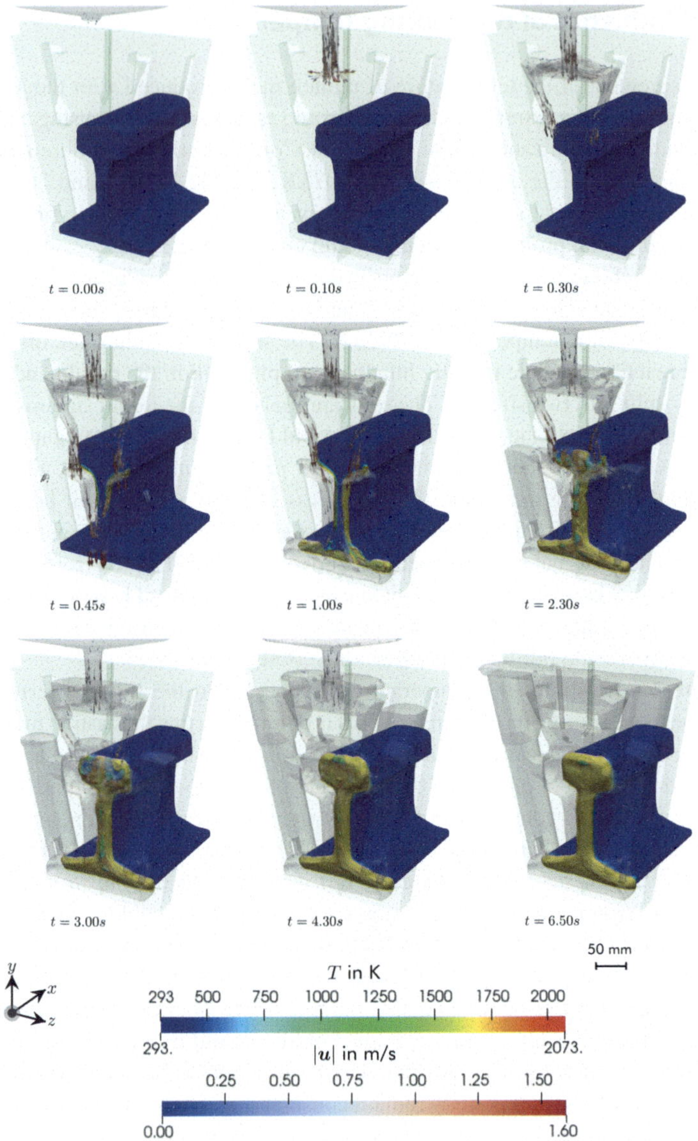

Figure 6.5 Velocity vectors in the liquid Thermit-steel and the temperature distribution depicting the melting of the rail during the pouring stage for the pouring temperature $T_\mathrm{P} = 2073\,\mathrm{K}$

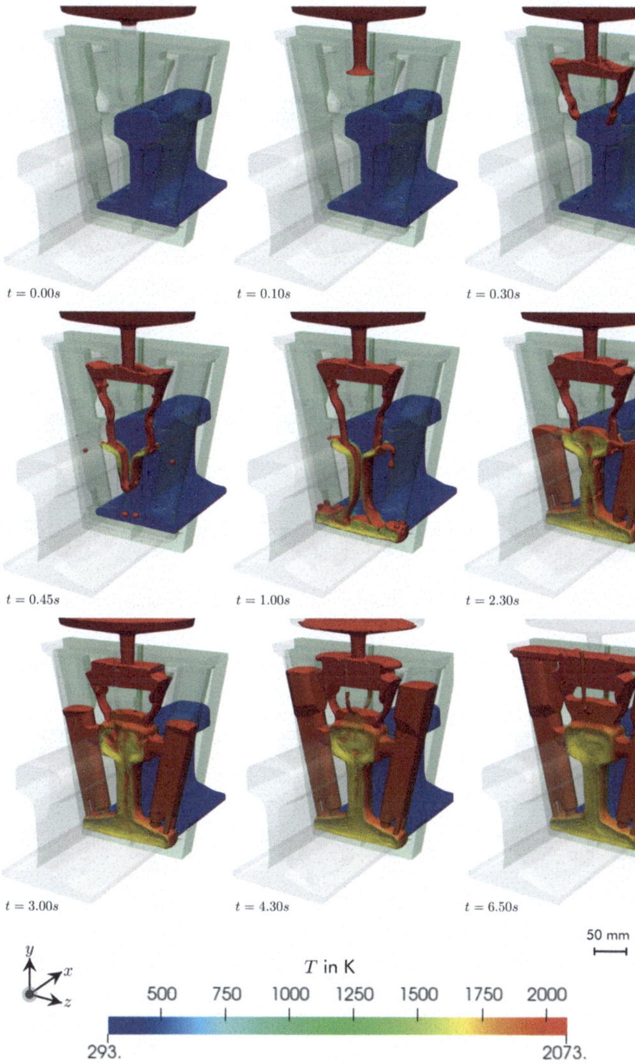

Figure 6.6 Temperature field of the liquid Thermit-steel during the pouring stage for the pouring temperature $T_P = 2073\,K$

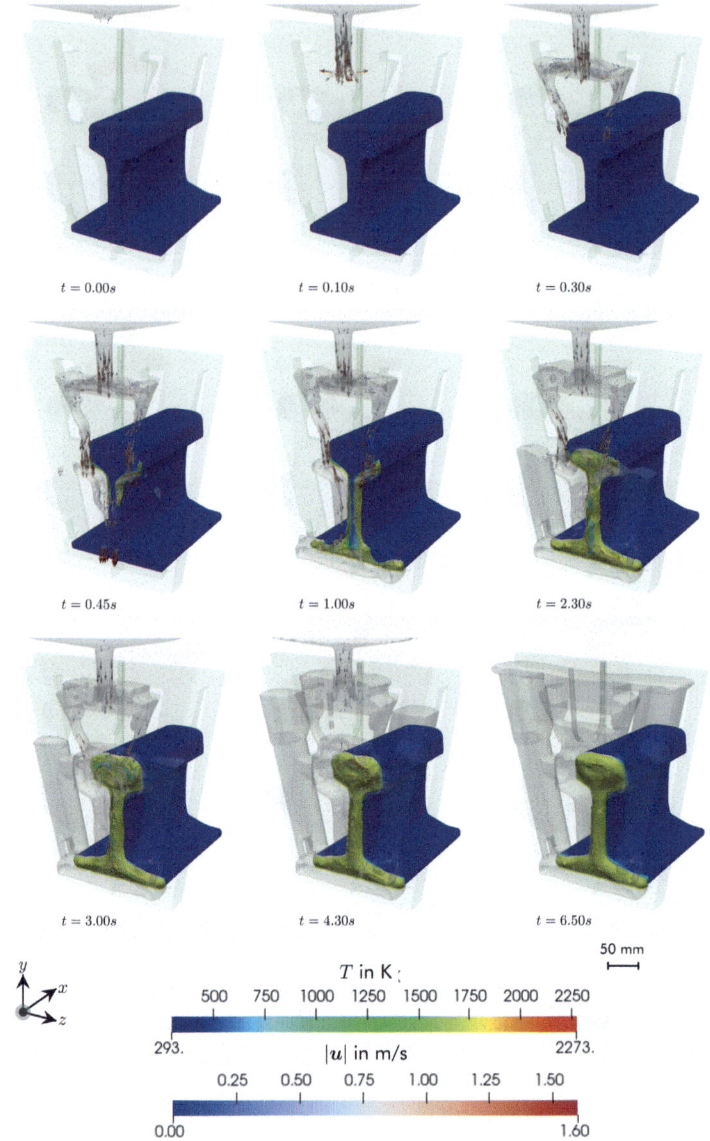

Figure 6.7 Velocity vectors in the liquid Thermit-steel and the temperature distribution depicting the melting off of the rail during the pouring stage for the pouring temperature $T_P = 2273\,K$

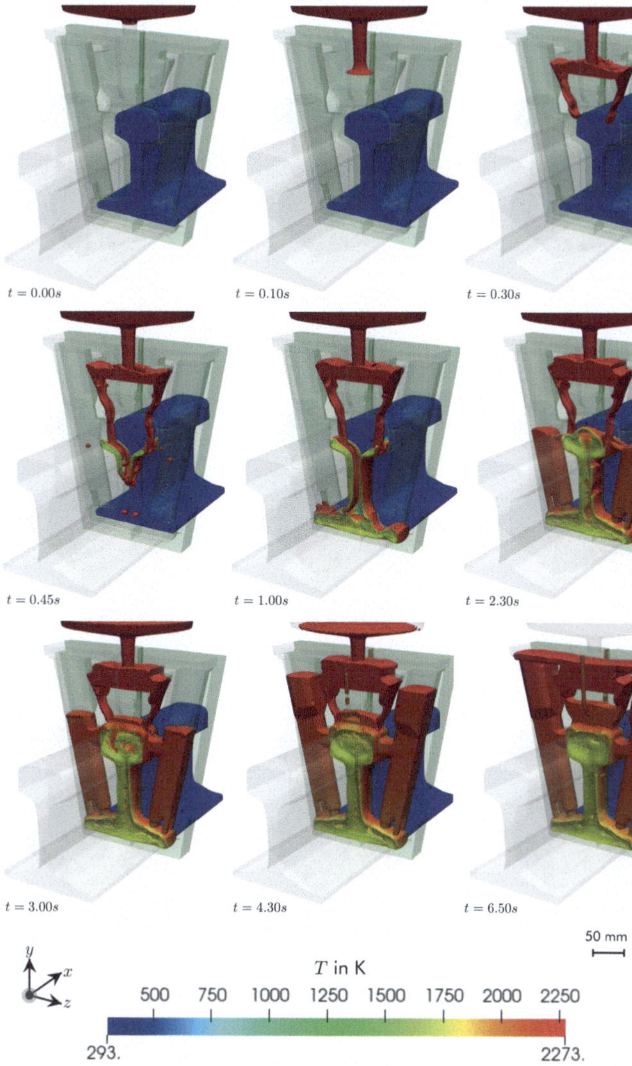

Figure 6.8 Temperature field of the liquid Thermit-steel during the pouring stage for the pouring temperature $T_P = 2273\,K$

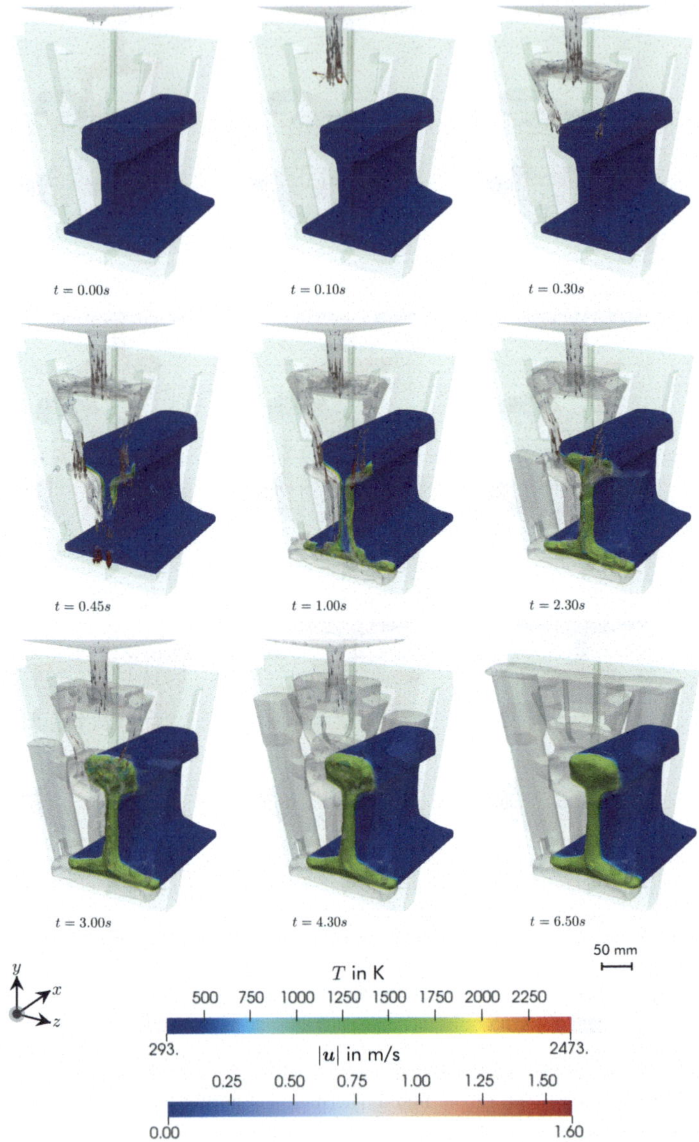

Figure 6.9 Velocity vectors in the liquid Thermit-steel and the temperature distribution depicting the melting off of the rail during the pouring stage for the pouring temperature $T_P = 2473\,K$

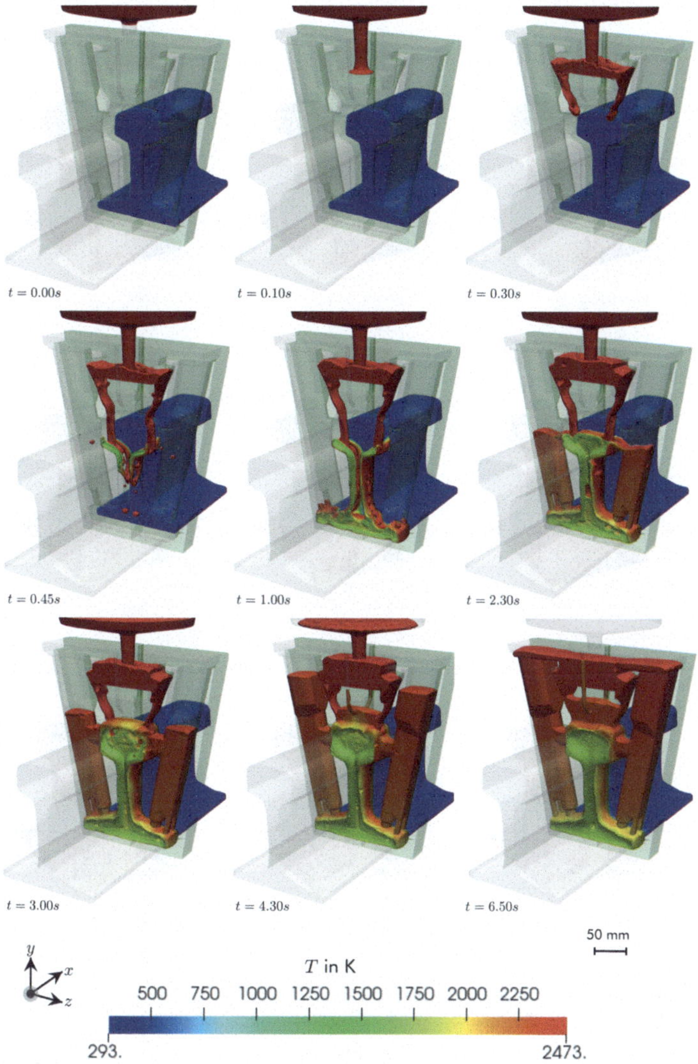

Figure 6.10 Temperature field of the liquid Thermit-steel during the pouring stage for the pouring temperature $T_P = 2473\,\mathrm{K}$

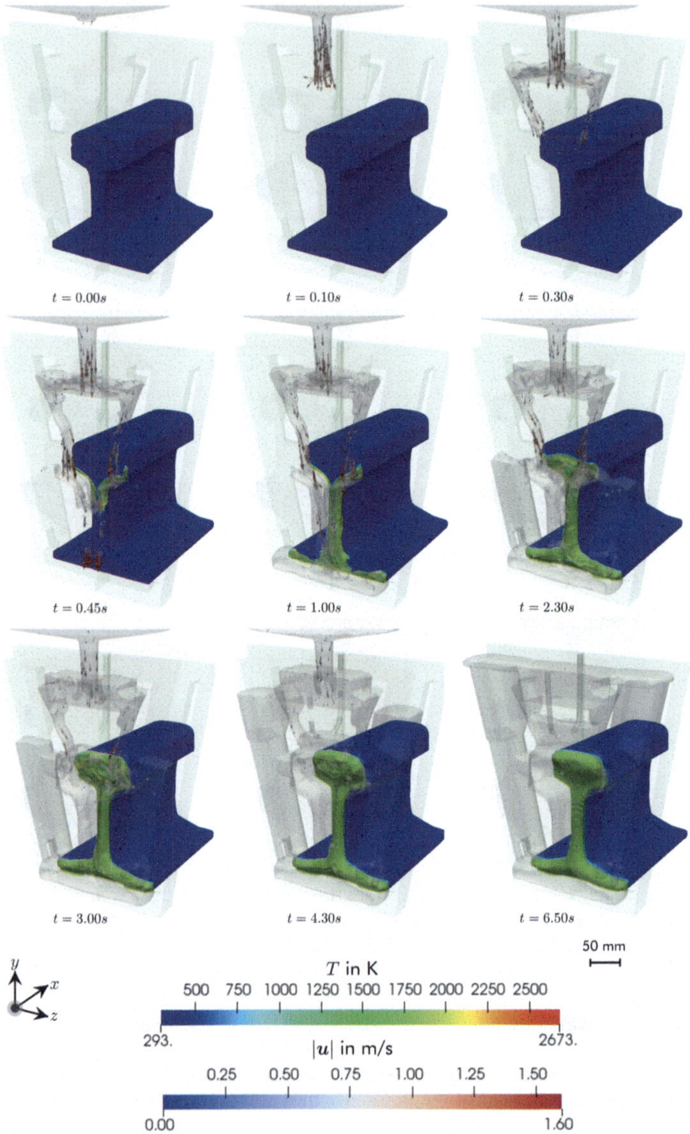

Figure 6.11 Velocity vectors in the liquid Thermit-steel and the temperature distribution depicting the melting off of the rail during the pouring stage for the pouring temperature $T_{\mathrm{P}} = 2673\,\mathrm{K}$

liquid Thermit-steel within the mould was below $|\boldsymbol{u}| = 1.6\,\mathrm{m\,s^{-1}}$ during the mould-filling process in all cases.

Subsequently, the temperature distribution in the liquid Thermit-steel during the pouring stage is also investigated for the remaining of the pouring temperatures and is illustrated in Figures 6.8, 6.10 and 6.12, respectively. It is observed that, with an increase in pouring temperature, there is also an increase in the amount of material melted off from the parent rail, resulting in a larger weld gap and a higher volume of liquid Thermit-steel being deposited in the weld gap.

Moreover, the peak temperatures of liquid Thermit-steel within the weld gap are also evaluated at the weld centre after the end of the pouring stage. It is found that the temperature of the liquid Thermit-steel increases from the rail base to rail head. However, it still remains well well below the initial pouring temperature, as is depicted in Figure 6.13. A qualitative analysis of the influence of different pouring temperatures on the shape of the melted rails within the mould after the pouring stage is presented in Figure 6.14. For the thermophysical properties employed in this study, the numerical results indicate that the melting of the parent rail during the pouring stage is extremely underestimated at the pouring temperature of 2073 K. In contrast, it is significantly overestimated at the pouring temperature of 2673 K. Pouring temperatures 2273 K and 2473 K exhibit acceptable agreement with the experiments. However, slight deviations are observed in the numerically obtained meted rails within the mould at the end of pouring stage compared to the experiment. In the simulations, the melting in the railhead is prolonged (indicated by blue circles in Figure 6.14). This is attributed to the entrapment of the air bubble i.e., the presence of the air pocket within the mould, immediately above the railhead, markedly affecting the heat transfer between the liquid Thermit-steel and the railhead in that region. Moreover, it is also found that the melting of the railhead-web intersection (indicated by black rectangles in Figure 6.14) is overestimated in the simulations. The observed phenomenon can be attributed to the following factors: Firstly, the simulations do not take into account for the presence of the casting mould. Consequently the temperature of the liquid streams is relatively higher before their initial contact with the railhead-web intersection, resulting in slightly increased melting at that specific region compared to experimental conditions. Secondly, in the AT welding process, the rails are slightly elevated before the mounting of the casting mould to compensate for the potential thermal stresses and shrinkage that may occur during the solidification stage. Consequently, the assumption of a uniform weld gap of 25 mm from rail base to railhead in the simulations, is subject to scrutiny.

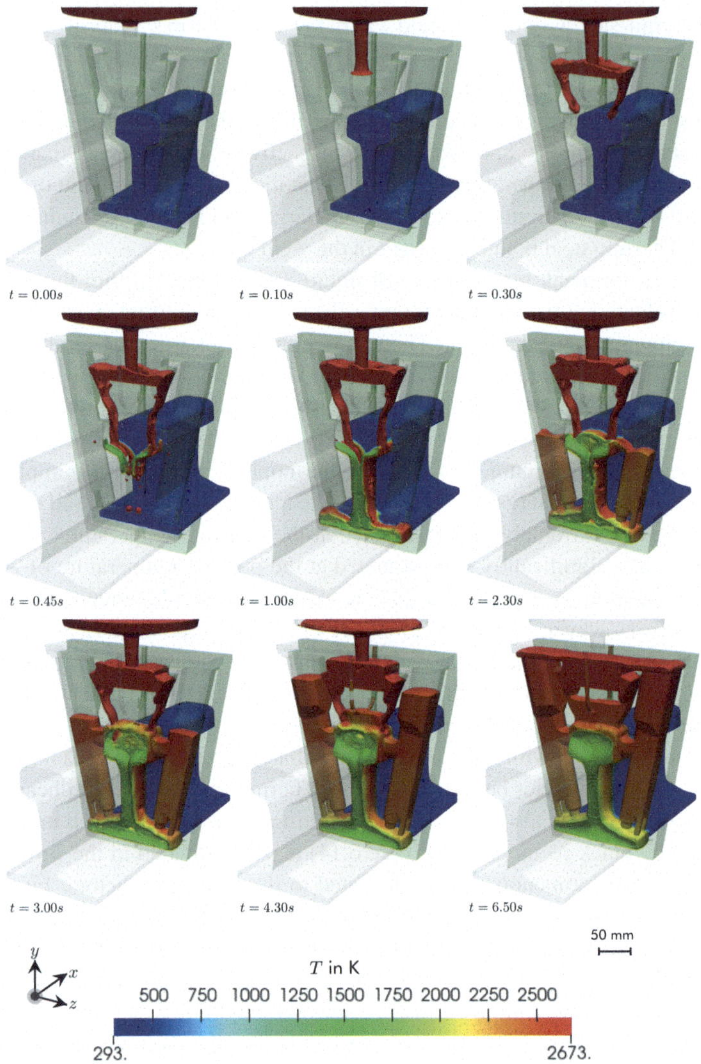

Figure 6.12 Temperature field of the liquid Thermit-steel during the pouring stage for the pouring temperature $T_P = 2673\,K$

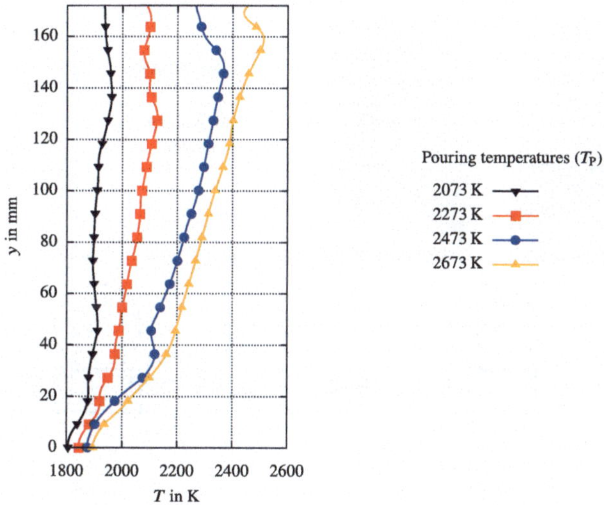

Figure 6.13 Peak temperatures at the weld center after the end of pouring stage (i.e., at t =6.50 s) for various pouring temperatures investigated in this study

The dimensions of the numerically obtained melted rails post-pouring stage at different pouring temperatures are compared with the experiment in Figure 6.15. The results are assessed on the vertical longitudinal plane (refer to Chap. 5). The rails within the mould before the pouring stage are depicted by the dashed vertical lines (black), while the melted rails are represented by curves (coloured lines with symbols) and are evaluated at solidus temperature (T_{Sol} = 1653.15 K). A vase-shaped weld deposit is observed for the pouring temperatures investigated in this study, with the exception of 2073 K, where the melted rails displayed a profile nearly parallel and coincident with the original rail shape, indicating insufficient melting compared to the original rail.

Furthermore, it is observed that the width of the weld gap between the melted rails, at the rail web and rail base (referred to as the region below y = 80 mm in Figure 6.15) is broader in experiments than in simulations. This disparity is likely attributed to the draining of the molten metal from the mould, resulting in further melting of the rails during the process.

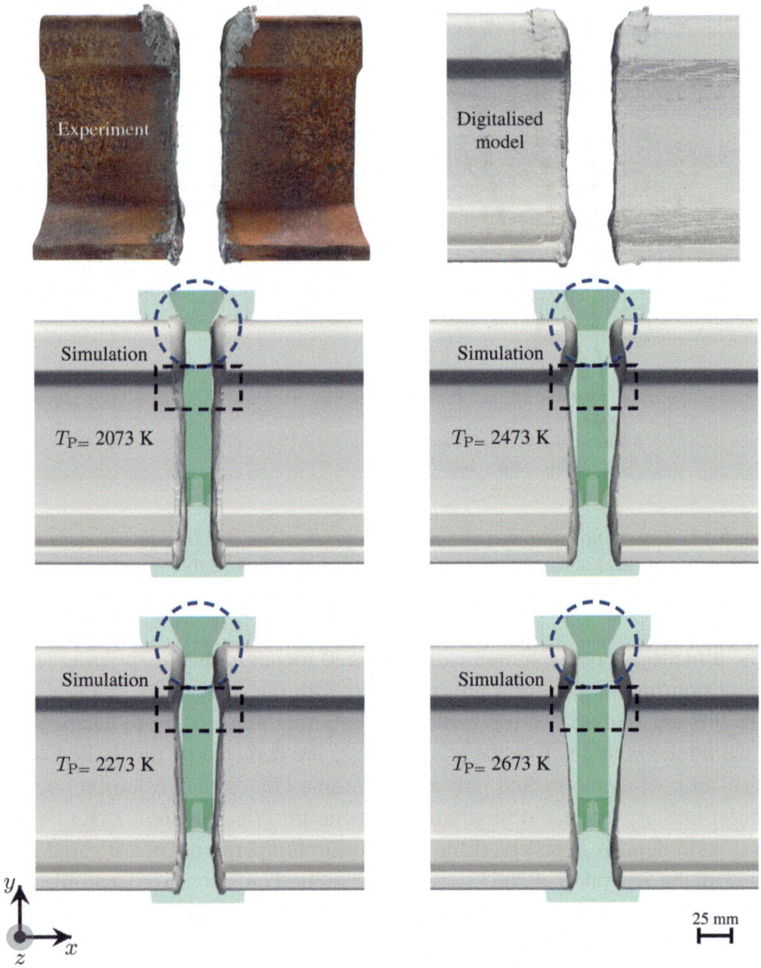

Figure 6.14 Qualitative analysis of the shape of the melted rail profiles within the mould post-pouring stage: Experiment (top left), photogrammetry (top right) and numerically obtained melted rail profile at various pouring temperatures—**Note:** The casting mould (transparent green) is shown only for better understanding

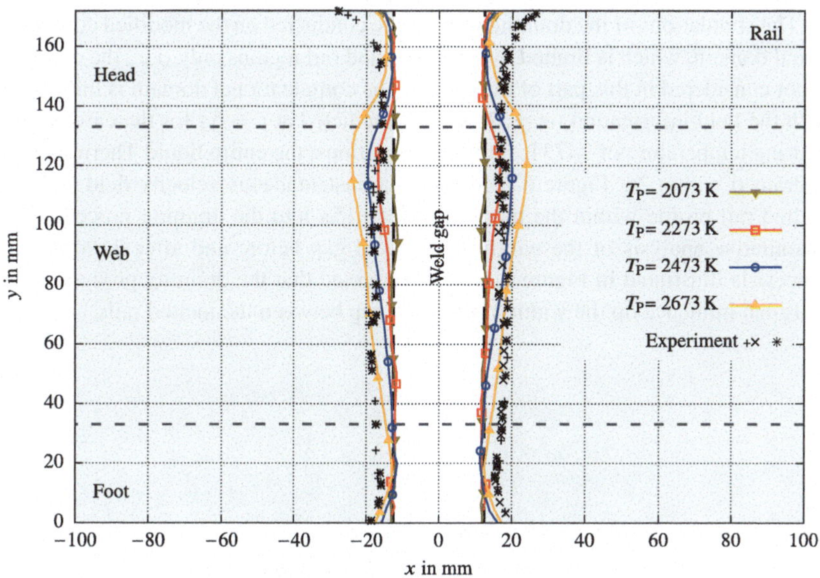

Figure 6.15 Comparison of the dimensions of the melted rails within the mould post-pouring stage from experiments and simulation for different pouring temperatures

6.2.4 Influence of the Draining Process

To assess the hypothesis positing that the width of the weld gap between the melted rail profiles is influenced by the draining of the molten metal, the simulation is extended for a scenario with pouring temperature of 2473 K. In this case, an outlet boundary condition is imposed on the bottom of the mould, and the mould already containing liquid Thermit-steel, undergoes the process of draining. As discussed in Chap. 5, the mould draining process is initiated as soon as the molten metal reaches the top of the mould during the filling process, which is achieved at approximately $t = 5$ s in the simulations.

The simulations of the draining process are conducted on the modified computational domain, which is limited to the mould and rail regions only (i.e., the crucible is not considered in this part of the study). The computational domain is initialised with the field information ($\alpha, \gamma, T, \boldsymbol{u}, \ldots$) predicted at $t = 5\,\text{s}$ for the case with a pouring temperature of 2473 K.[3] In the simulations, the entire liquid Thermit-steel is drained within 2 s. Figure 6.16 depicts the instantaneous velocity field and the melted rail profile within the mould at $t = 0.15\,\text{s}$ into the draining process. The quantitative analysis of the width of the weld gap before and after the draining process is illustrated in Figure 6.17. It is observed that the draining process has a marginal influence on the width of the weld gap between the melted rails.

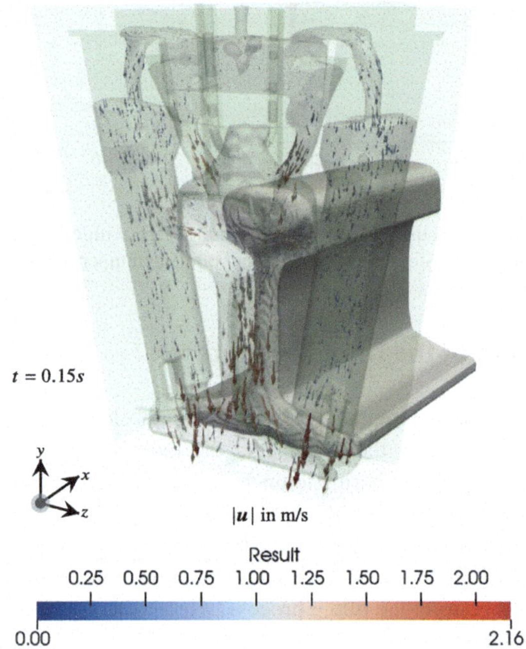

Figure 6.16 Instantaneous velocity vector field in the liquid Thermit-steel and the melted rail within the mould during the draining process at $t = 0.15\,\text{s}$ for case with a pouring temperature $T_P = 2473\,\text{K}$

[3] For this purpose `mapFields` utility in OpenFOAM is used.

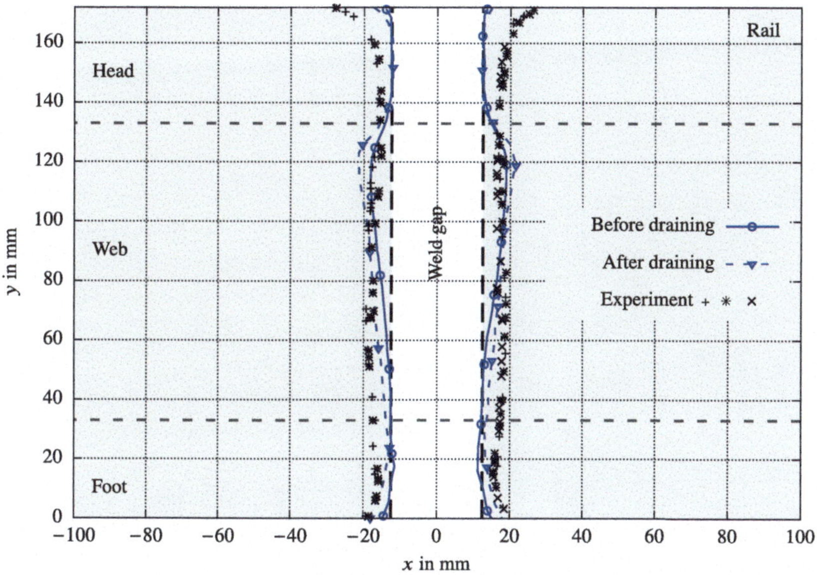

Figure 6.17 Comparison of the width of the weld gap between the melted rails before (at $t = 5$ s) and after draining the liquid Thermit-steel in the simulation at a pouring temperature $T_P = 2473$ K along with experimental data

* * *

In conclusion, the focus of this chapter was on simulating the thermal pouring stage of the rapid welding without preheating (RWOP) process. The turbulent characteristics of the flow, arising from both the discharge process and the complex shape of the mould during filling, were described using the $k - \varepsilon$ turbulence model. The estimated Reynolds number ($Re = uD/\nu$) at the outlet of the crucible was determined to be approximately 4.5×10^4. Following a grid independent study, a grid resolution of approximately 7.78 million cells was selected for the simulations, which were executed on the HPC-Cluster of the TU Bergakademie Freiberg using 256 cores, resulting in significant computational expense. Moreover, the influence of various pouring temperatures on the shape of the melted rail profile post pouring stage was investigated, revealing a significant impact. Discrepancies between the experimentally obtained shape of the melted rails and those obtained numerically were discussed in detail. It is important to note that the comparison of simulation results was constrained to only one set of experimental data, raising concerns about the accuracy of validating the numerical investigations solely on this basis.

Simulation of the Solidification Process 7

As previously demonstrated, the simulation of turbulent flow during the thermal pouring stage (involving crucible discharge and mould-filling) in conjunction with solid-liquid phase change phenomena is deemed computationally expensive. Therefore, in this chapter, specific process assumptions are employed to simplify the complexity of the problem at hand, aiming to avoid the simulation of the pouring stage and markedly reduce the computational demands [13, 20]. Nevertheless, in this study, the following key assumptions are made: Firstly, the casting mould and the sand-plug are not physically considered in the simulations, instead, they are implemented as an adiabatic boundary condition. Secondly, the mould is pre-filled with molten metal, maintaining a constant initial molten metal temperature. Finally, the rail profile is simplified to a flat surface, assuming that the rail remains solid and is not subjected to melting during the mould-filling stage. Through the incorporation of these assumptions, not only are the governing equations simplified to laminar flow conditions but also the results, such as the temporal and spatial evolution of the solid-liquid interface front, maximum width of the FZ and HAZ, can be achieved and predicted in a significantly shorter span of time. It is noteworthy that some of the results presented in this chapter have been published in Kewalramani et al. [97] and the findings are built upon and extended in the current study, offering additional insights.

7.1 Simulation Set-up

Mesh generation

A mesh generation strategy similar to that discussed in Sect. 6.1 is employed. The cell size within the mould, depicted in the orange region of Figure 6.1(b), is refined

R. G. Kewalramani, *Computational Thermo-Fluid Dynamics of Aluminothermic Welding Process*, https://doi.org/10.1007/978-3-658-47535-2_7

compared to the cell size in the rail, depicted in the red region. Given the exclusion of the pouring stage in this study, the crucible is not considered in the simulations. Finally, to optimise computational efficiency, only one-quarter of the computational domain (as depicted in Figure 7.1) is modelled by taking advantage of the bilateral symmetry of the welding setup.

Initial and boundary conditions

Figure 7.1 illustrates the initialisation of the fields at time $t = 0\,\mathrm{s}$. It is assumed that the mould is pre-filled and the initial velocity of the molten metal within the mould is $\boldsymbol{u} = 0\,\mathrm{m\,s^{-1}}$. The steel-slag ratio inside the mould is taken from the simulations performed by Weiss et al. [16]. The temperature of the molten metal along the y-axis within the mould is set to a constant value $T_{\mathrm{MM}} = 2473\,\mathrm{K}$, while the temperature of the rail is set to the ambient temperature $T_{\mathrm{R}} = T_{\infty} = 293\,\mathrm{K}$.

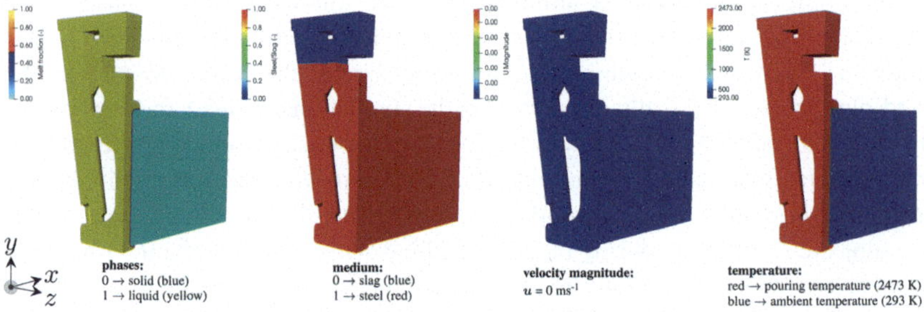

Figure 7.1 Initialisation of fields: melt fraction, volume fraction, velocity and temperature field at time $t = 0$ s [97]

The exploded view of the computational domain is presented in Figure 7.2 for a more convenient discussion of the thermal boundary conditions. Taking advantage of two symmetry planes (along the longitudinal length of the rail and across the weld centre), only a quarter of the casting system is considered for the simulations.

- At the **top** boundary, the heat loss into the surroundings occurs through both radiation and convection. Consequently, a convective-radiative boundary condition is applied:

$$\kappa \frac{\partial T}{\partial n} = -h_{\mathrm{c}}(T - T_{\infty}) - \varepsilon_{\mathrm{r}}\sigma_{\mathrm{r}}(T^4 - T_{\infty}^4), \tag{7.1}$$

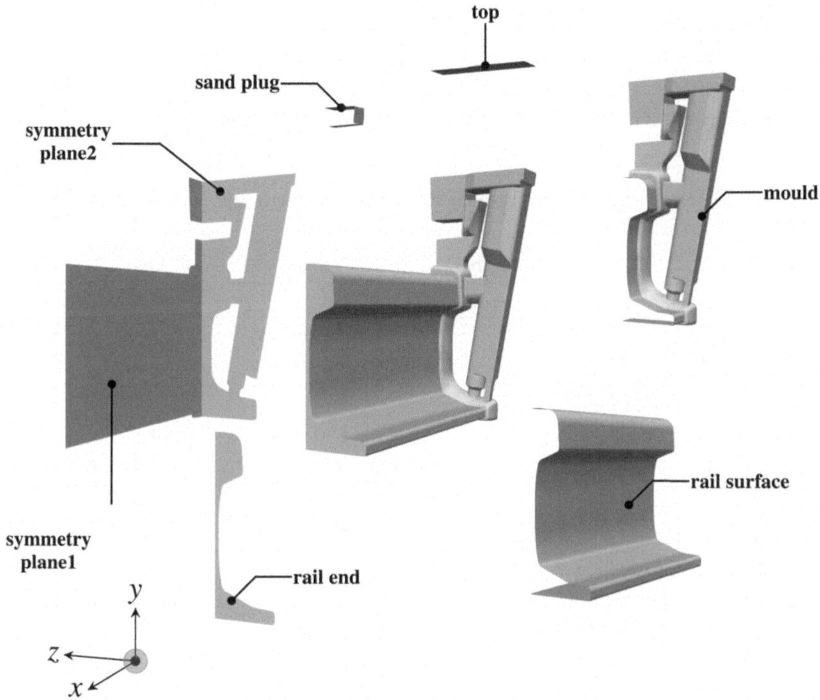

Figure 7.2 An exploded view of all the boundaries of the computational domain [97]

where h_c represents the convective heat transfer coefficient, ε_r is the emissivity coefficient of the molten metal, σ_r is the Stefan-Boltzmann constant and T_∞ is the ambient temperature.

- At the **rail surface** boundary, the heat loss into the surrounding is due to convection:

$$\kappa \frac{\partial T}{\partial n} = -h_c(T - T_\infty). \tag{7.2}$$

- Adiabatic wall conditions are applied to the **mould**, **sand plug** and **rail end** boundaries:

$$\frac{\partial T}{\partial n} = 0. \tag{7.3}$$

- Finally, symmetry boundary conditions are imposed on the two symmetry planes.

Material properties

The thermophysical properties of the materials, Thermit-steel (liquid), rail (solid) and slag employed in the CFD-Simulations are given in Table 7.1.[1]

Table 7.1 Material properties employed in the simulations for liquid Thermit-steel, rail steel and slag [9, 13, 16, 101]

Property	Symbol	Value	Unit
	Thermit-steel and rail		
Density	ϱ	6900	kg m^{-3}
Dynamic viscosity	μ	5×10^{-3}	kg m^{-1} s
Specific heat capacity, solid	$c_{p,\mathrm{sol}}$	640	J kg^{-1} K^{-1}
Specific heat capacity, liquid	$c_{p,\mathrm{liq}}$	680	J kg^{-1} K^{-1}
Thermal conductivity, solid	κ_{sol}	30	W m^{-1} K^{-1}
Thermal conductivity, liquid	κ_{liq}	55	W m^{-1} K^{-1}
Latent heat of fusion	$\mathcal{L}$	277×10^3	J kg^{-1}
Thermal expansion coefficient	β	1.2×10^{-4}	K^{-1}
Solidus Temperature	T_{sol}	1653	K
Liquidus Temperature	T_{liq}	1738	K
	slag		
Density	ϱ	3740	kg m^{-3}
Dynamic viscosity	μ	0.2273	kg m^{-1} s
Specific heat capacity	c_p	1000	J kg^{-1} K
Thermal conductivity	κ	10	W m^{-1} K

Numerical parameters

The numerical schemes utilised for the simulations are summarised in Table 7.2. Furthermore, the time step is determined by a maximum Courant number (Co) below 0.75, leading to a typical time step of the order $\Delta t = 10^{-4}$ s. Finally, the convergence criteria are satisfied when the normalised residual values of the continuity, momentum and energy equations are below $\times 10^{-8}$.

[1] For the sake of completeness, the material properties of the liquid Thermit-steel and rail, as listed in Table 6.1 are reiterated in Table 7.1.

Table 7.2 Numerical discretisation schemes

Term	Operator	Scheme	OpenFOAM
Temporal	$\frac{\partial \varphi}{\partial t}$	Euler	`Euler`
Gradient	$\nabla \varphi$	Central differencing	`Gauss linear`
Divergence	$\nabla \cdot (\varrho \boldsymbol{u}\boldsymbol{u})$	Linear Upwind	`Gauss linearUpwindV`
	$\nabla \cdot (\varrho c_p \boldsymbol{u} T)$	Limited linear differencing	`Gauss limitedLinear 1`
	$\nabla \cdot (\gamma \boldsymbol{u})$	van Leer limiter	`Gauss vanLeer01`
	$\nabla \cdot (\alpha \overline{\boldsymbol{u}})$	van Leer limiter	`Gauss vanLeer01`
	$\nabla \cdot (\alpha \boldsymbol{u}_{\mathrm{r}})$	Interface compression	`Gauss interfaceCompression`
Laplacian	$\nabla \cdot (\nabla \varphi)$	Central differencing	`Gauss linear corrected`
		Explicit non-ortho. correction	

7.2 Results and Discussions

Herein, the RWOP process of the AT welding of rails is investigated to identify the suitable simulation parameters. Furthermore, the temporal evolution of the solid-liquid interface front and the flow field within the mould during the solidification stage are discussed in detail. Additionally, the influence of various parameters such as the initial temperature of the molten metal and the thermophysical properties of Thermit-steel and rail on the final weld profile is also explored.

7.2.1 Mesh Sensitivity Analysis

A mesh sensitivity analysis is performed with four different cell sizes, i.e., $\Delta x = 1.50$ mm, 1.00 mm, 0.75 mm and 0.50 mm, respectively, to determine the influence of the computational mesh on the solid-liquid interface front. The size of the computational domain for each cell size is listed in the Table 7.3. The simulations are carried out for the molten metal temperature of $T_{\mathrm{MM}} = 2473$ K. The position of the solid-liquid interface front, evaluated on the symmetry plane (along the longitudinal length of the rail) for various cell sizes at $t = 10$ s, are presented in Figure 7.3.

Table 7.3 Size of the computational domain

Cell size	cells
mm	$\times 10^6$
1.50	0.43
1.00	0.79
0.75	1.75
0.50	5.26

As depicted in Figure 7.3, no significant change is observed in the solid-liquid interface front beyond the cell size of $\Delta x = 0.75$ mm. Therefore, a computational mesh comprising 1.75 million cells with $\Delta x = 0.75$ mm is selected for further investigations.

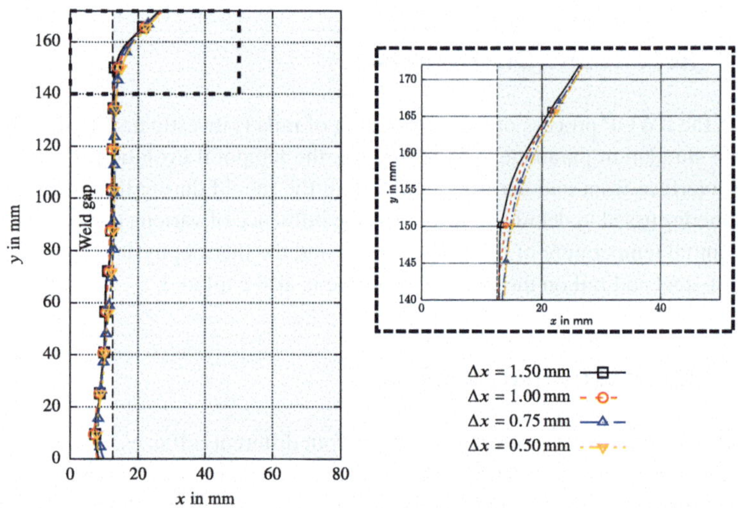

Figure 7.3 Position of the solid-liquid interface front for different cell sizes at time $t = 10$ s

7.2.2 Melting/Solidification

Figure 7.4 shows the solid-liquid interface front at $t = 2\,s$, $10\,s$ and $40\,s$, respectively, during the cooling stage of the AT welding process, as predicted from the 3D axis symmetrical simulation. For better visualisation, the computational domain is mirrored to represent the results on the entire rail geometry. The melt fraction in steel ranges from 0 in the solid phase ($T \leq T_{sol}$, grey) to 1 in the liquid phase ($T \geq T_{liq}$, yellow) and the 3D solid-liquid phase front is evaluated at $T_{sol} = 1653\,K$. As stated at the beginning of this chapter, the mould is pre-filled with the molten metal temperature of 2473 K, and the temperature of the rail is initialised to 293 K. It is observed that the rail end inside the mould starts melting immediately within the first few seconds. This is caused by high-temperature gradients during the initial stages of heat transfer from the molten metal to the rail end resulting in thermal convection.

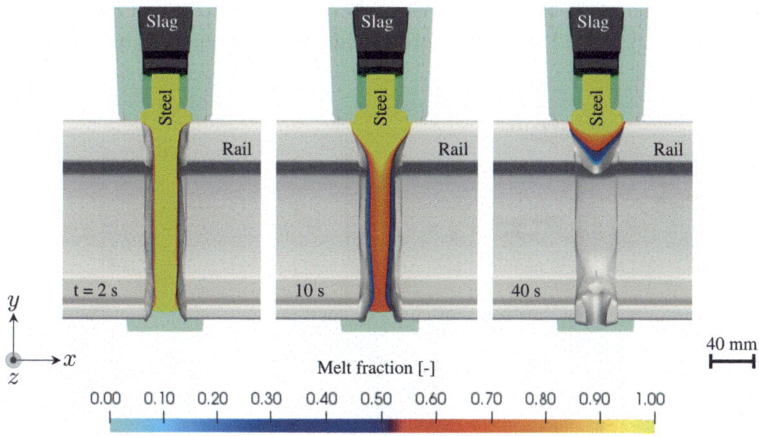

Figure 7.4 The solid-liquid interface front for the initial molten metal temperature of 2473 K at $t = 2\,s$, $10\,s$ and $40\,s$

The primary heat flux inside the mould goes through the rail end. Consequently, the temperature of the molten metal drops in the vicinity of the rail end. Thus, at around $t = 10\,s$ into the cooling process, the molten metal in the weld gap (from rail base to rail web) is in a mushy state. However, at the edge of the rail base, the molten metal still exists in the mushy state at around $t = 40\,s$, even after the centre of the rail base and rail web regions are completely solidified. This is because the temperature of the bulk molten metal is higher in the foot-side-riser. Moreover, it is evident that the melting of the railhead is prolonged and requires the longest time to solidify. This is attributed to the bulk molten metal situated just above the railhead inside the mould and thermal convection. As a result, the solid-liquid interface front in the weld gap progresses inward from the rail end towards the weld centre and upward from the rail base towards the railhead, forming a V-shaped solidification pattern.

7.2.3 Flow Field Inside the Mould

The velocity field during the cooling stage at different times is illustrated in Figure 7.5. At around $t = 0.5\,\text{s}$, the molten metal near the rail end descends along the cold solid-liquid interface front towards the rail base. This is caused by the rapid cooling of the molten metal near the rail end, leading to an increase in the density of the molten metal. Subsequently, the molten metal near the rail base, which is relatively cold, is heated up at the core (weld centre), rises upwards through the

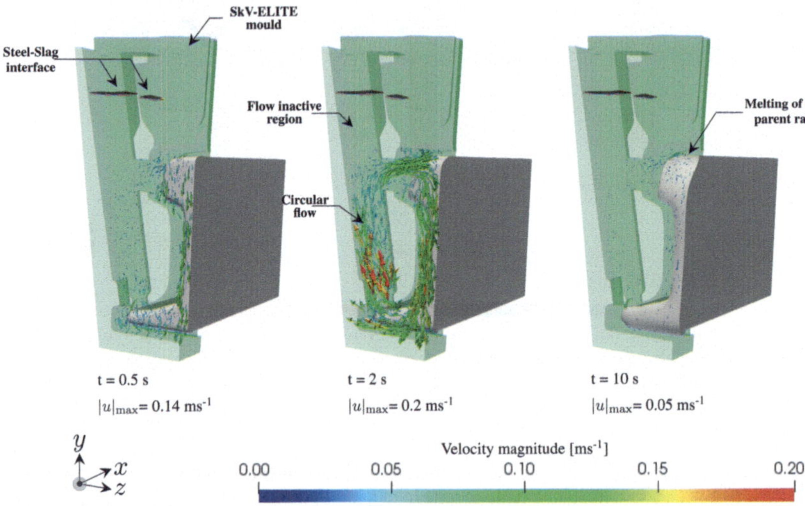

Figure 7.5 Flow field inside the mould during the initial stages of cooling: velocity vectors (coloured), mould (transparent green), 3D solid liquid interface profile and parent rail (solid grey) [97]

foot-side-riser and travels along the head-side-riser into the weld gap, where heat is released at the railhead. Therefore, the flow pattern and the solid-liquid phase change at the beginning of the cooling stage are governed by thermal convection. As the solidification progresses, the mushy region is built up in the weld gap (refer to Figure 7.8, $t = 10\,\text{s}$, red-blue region), causing a decrease in the velocity magnitude of the convective flow. From around $t = 20\,\text{s}$, it can be inferred that heat transfer between the molten metal and the rail end inside the mould is mainly through heat conduction.

7.2.4 Influence of the Initial Molten Metal Temperature

In the preceding section, the results are discussed based on the initial molten metal temperature of 2473 K. However, as already mentioned in Sect. 6.2.3 measuring the temperature of the molten metal after the Thermit® reaction and during discharge from the crucible proves to be very challenging. Therefore, a systematic study is performed to explore the influence of four different molten metal temperatures—$T_{\text{MM}} = 2073\,\text{K}, 2273\,\text{K}, 2473\,\text{K}, 2673\,\text{K}$ on the FZ and HAZ.

At the molten metal temperature of 2073 K (Figure 7.6), the evolution of the solid-liquid interface front is observed to be dominant from the rail end to the weld centre (inwards, i.e., in the horizontal direction). Due to the large mass and contact area with the molten metal in the railhead and rail base (centre) regions, the solidification progresses faster, resulting in an isolated melt in the rail web region (around $t = 13\,\text{s}$). Consequently, the volume deficit resulting from the solidification contraction cannot be compensated by liquid metal as the liquid feeding path from the molten metal reservoir located above the railhead or through the head-side-riser in the rail web region is completely blocked. This phenomenon gives rise to *shrinkage defects*[2]. Furthermore, the weld gap in the rail region is completely solidified within $t = 40\,\text{s}$. Similar behaviour is also observed at the molten metal temperature of 2273 K (Figure 7.7). However, the solidification time is somewhat longer compared to the molten metal temperature of 2073 K. Nevertheless, in both cases, the maximum width of the FZ is smaller than the weld gap (25 mm). Thus, violating the European standards required for the approval of a weld [104], which stipulates that the minimum melting depth of the rail end inside the mould should be at least 3 mm.

[2] The developed numerical model does not explicitly consider volume expansion or contraction for the sake of simplicity. However, it is widely recognised that liquid steel contracts during solidification. In regions experiencing volume contraction, an increased risk of porosity formation exists due to insufficient liquid feeding during the solidification process [13].

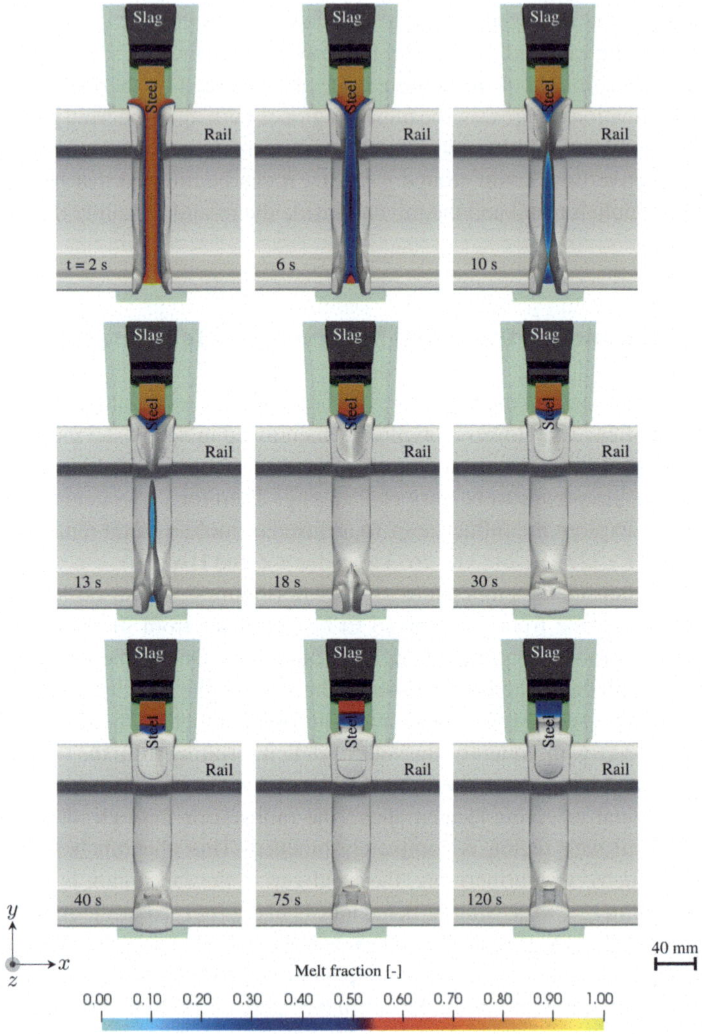

Figure 7.6 Temporal and spatial evolution of the solid-liquid interface front at the initial molten metal temperature of 2073 K

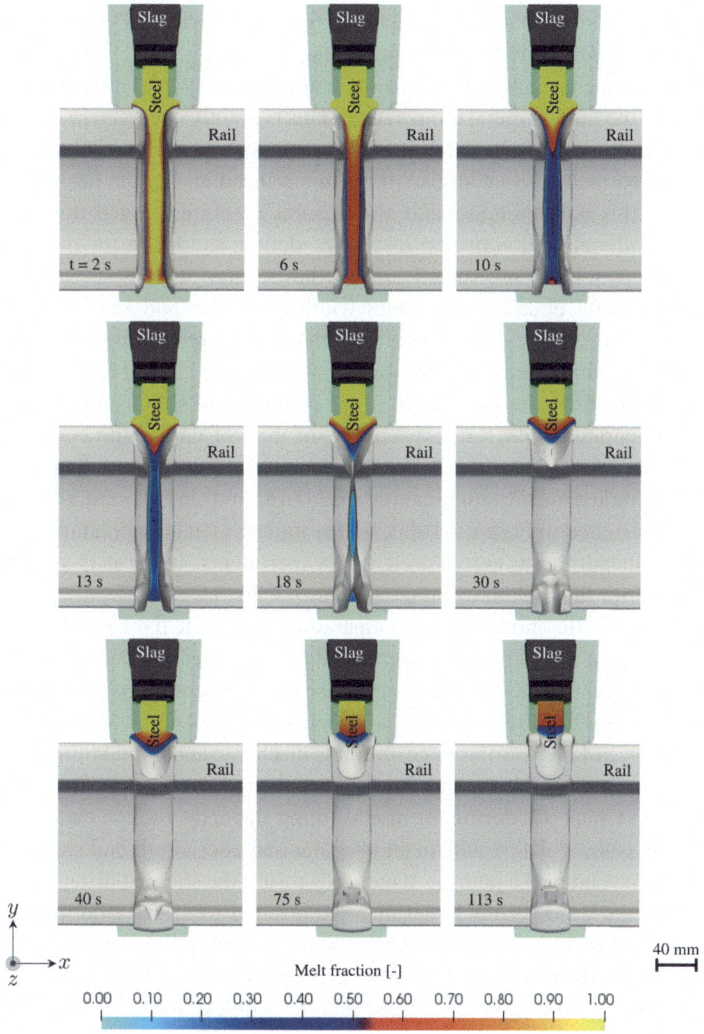

Figure 7.7 Temporal and spatial evolution of the solid-liquid interface front at the initial molten metal temperature of 2273 K

At molten metal temperatures of 2473 K (Figure 7.8) and 2673 K (Figure 7.9) the formation of the isolated liquid region in the web is either delayed or prevented. It is evident that the penetration of the molten metal into the solid rail increases with an increase in the molten metal temperature. This is mainly observed in the railhead area, due to the more prolonged thermal convection in the molten metal reservoir (or the bulk molten metal) situated just above the railhead and in the head-side-riser. Additionally, this bulk molten metal also ensures an uninterrupted flow of liquid metal in the rail web region in case of volume deficit, as long as the molten metal in the feeding path is not obstructed due to solidification.

The numerically obtained maximum width of the FZ and HAZ for different molten metal temperatures are compared with experiments at various positions in Table 7.4. In the simulations, the maximum width of the FZ and HAZ is obtained by evaluating the furthermost location of the isotherms of solidus temperature and eutectoid temperature (996 K), respectively, in the rail during the cooling stage. Figure 7.10 shows a comparison between the simulated maximum width of the FZ and HAZ at molten metal temperatures of 2473 K and 2673 K and experimental results. As illustrated in Figure 7.10(a), for the molten metal temperature of 2473 K the FZ in railhead region agrees well with experiments, while it is significantly overestimated for the molten metal temperature of 2673 K.

Conversely, for the molten metal temperature of 2673 K the FZ below the railhead region agrees well with experiments, but it is significantly underestimated for 2473 K. Although, the FZ obtained in the case of 2673 K, show a good agreement from below the railhead region with the experiments, it is noteworthy to mention, that the influence of the pouring stage (crucible discharge and mould filling) stage on the initial rail profile inside the mould is not considered. Nevertheless, as discussed in the previous Chap. (6), during the mould-filling stage, the molten metal melts the rail end in its path, which results in an irregular and altered rail end surface inside the mould at the end of the pouring stage. As a result, the shortest distance between the rails (i.e., the weld gap) is larger than the original weld gap, and more molten metal is settled inside the weld gap. Therefore, it is essential to consider the melted rail profile inside the mould to predict the maximum width of the FZ and HAZ. As shown in Figure 7.10(b) it is to be noted that the width of the HAZ for the molten metal temperature of 2473 K demonstrates a good agreement with experiments, whereas it is overestimated for the molten metal temperature of 2673 K. Additionally, the numerically obtained temperature distribution in the rail is compared with those from experiment at various locations during the cooling stage, as depicted in Figure 7.11.

It is evident that, the temperature distribution in the rail is significantly overestimated for the case with the molten metal temperature of 2673 K. Conversely,

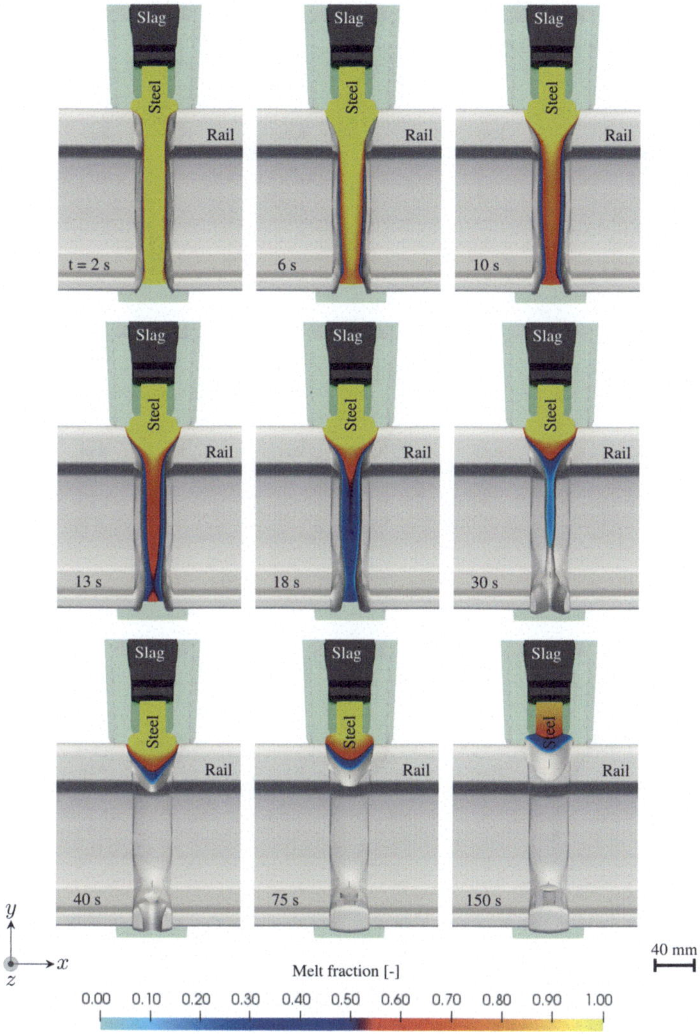

Figure 7.8 Temporal and spatial evolution of the solid-liquid interface front at the initial molten metal temperature of 2473 K

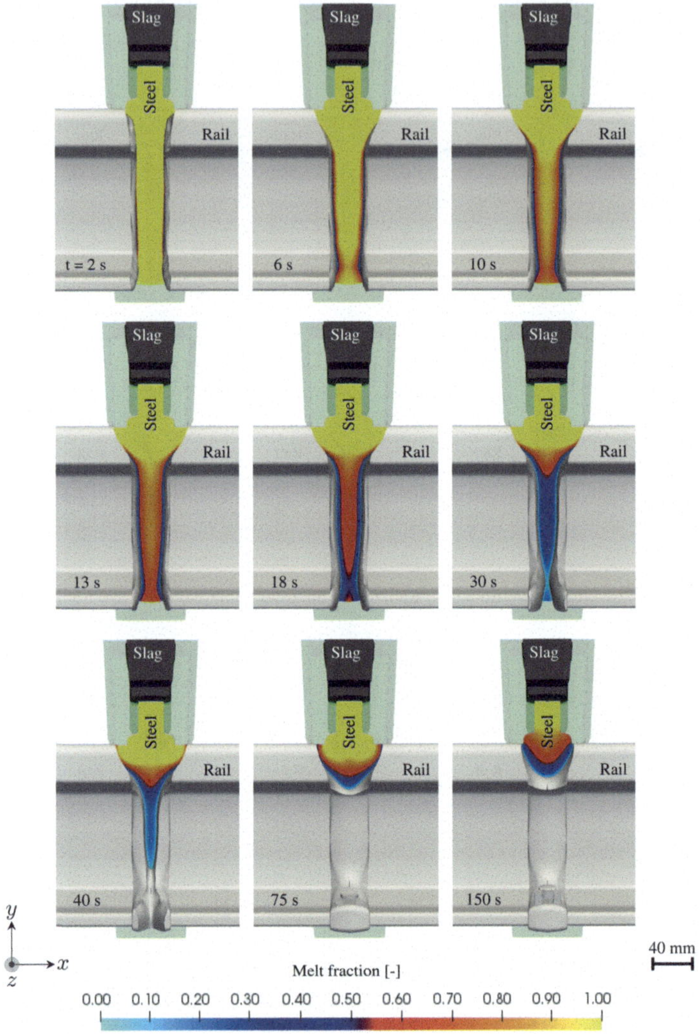

Figure 7.9 Temporal and spatial evolution of the solid-liquid interface front at the initial molten metal temperature of 2673 K

Table 7.4 Measurements of the FZ (A–E) and HAZ (F–H): Comparison between experimental data and simulation results for different molten metal temperatures (refer to Figure 5.11); published in [97]—**Note:** The grey colour in the table indicates melting below the required 3 mm limit

		FZ					HAZ		
		A	B	C	D	E	F	G	H
		mm	mm	mm	mm	mm	mm	mm	mm
Exp.	Weld 1	65	32	34	28	39	84	55	55
	Weld 2	69	34	33	32	39	85	53	55
Sim.(Molten metal Temperature)	2073 K	25	25	25	25	39	55	42	41
	2273 K	49	25	25	25	39	69	47	45
	2473 K	62	33	28	25	39	89	56	50
	2673 K	86	40	34	27	39	110	65	55

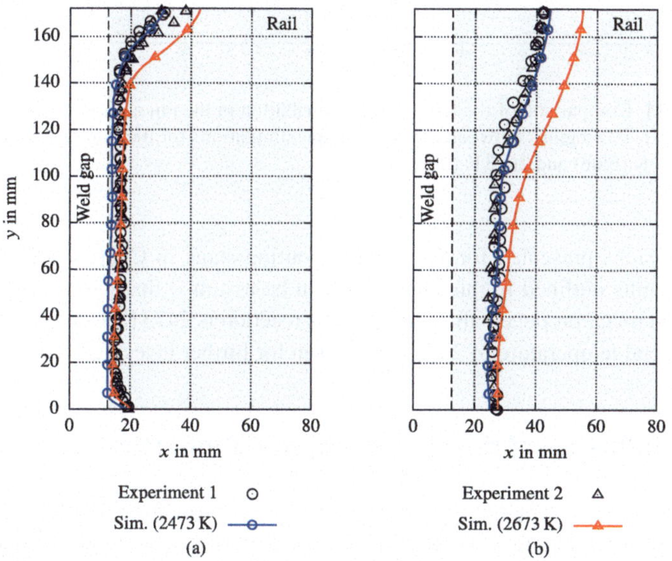

Figure 7.10 Comparison of the maximum width of the FZ (a) and HAZ (b) between experiments and simulations for the molten metal temperatures 2473 K (blue) and 2673 K (red)

the numerical predictions for the case with the molten metal temperature of 2473 K show a satisfactory agreement with the experimental data. Therefore, considering

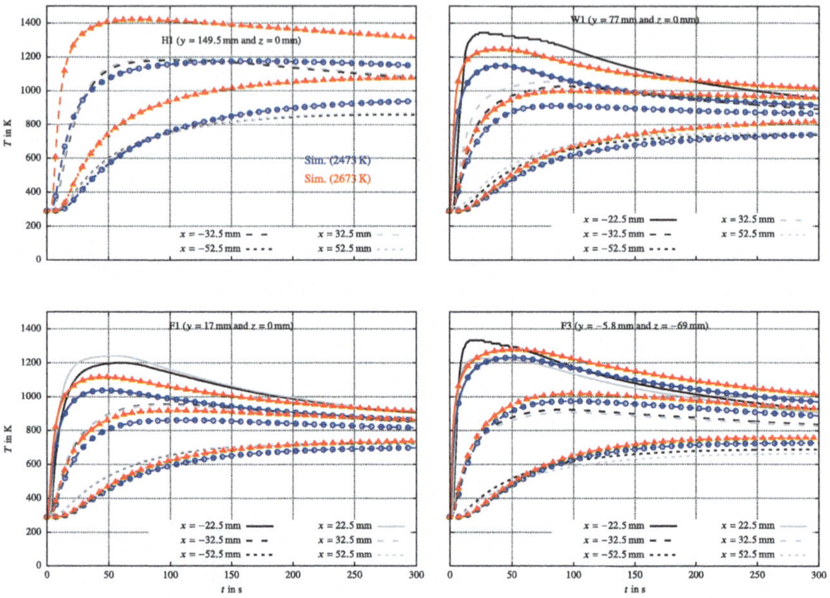

Figure 7.11 Comparison of the temperature distribution in the rail at head (H1), web (W1) and base (F1, F3) regions between experiment and simulations for the molten metal temperatures 2473 K (blue) and 2673 K (red)

the discussions presented for the thermal pouring stage in Chap. 6 and the simulation results outlined in this chapter, it can be assumed that the temperature of the molten metal produced after the Thermit-reaction is 2473 K. Consequently, the molten metal temperature of 2473 K is chosen for further investigations.

7.2.5 Influence of the Thermophysical Properties

In this thesis, the thermophysical properties utilised to predict the flow field inside the mould, the temporal evolution of the solidification front, and the maximum width of FZ and HAZ are considered constant and assumed to be independent of temperature. Moreover, measurements of the thermophysical properties of Thermit-steel at temperatures beyond the liquidus temperature is an extremely challenging task, and such data for the liquid Thermit-steel is not completely available in the literature. As a consequence, values for pure iron (or compositions of steel that resemble Thermit-steel) that have been already documented in the literature are employed. Therefore, it is essential to understand the influence of various values of thermophysical properties of the liquid Thermit-steel on the maximum width of the FZ. This section delves into the systematic study of how different (constant) values of density, viscosity, thermal conductivity and specific heat capacity, as used in the literature, impact the maximum width of the FZ.

Density

The numerically obtained maximum width of the FZ for three distinct cases (namely, V_1, V_2 and V_3) involving different density values for solid and liquid phases of steel is compared with the data from experiments and reference simulation[3] as shown in Figure 7.12. In V_1 the density of the solid phase is changed from 6900 kg m^{-3} in the reference simulation to 7500 kg m^{-3} [22], while in V_2, only the density of liquid phase is changed from 6900 kg m^{-3} to 6420 kg m^{-3} [18, 105].

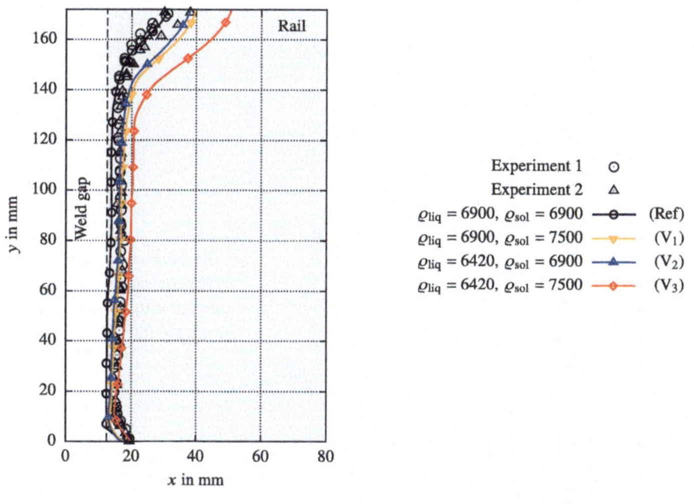

Figure 7.12 Comparison of the maximum width of the FZ on the symmetry plane along the x-axis between: Experiments (symbols) versus reference simulation (black solid line with symbol) with various density (solid, liquid) values in kg m^{-3} (coloured solid lines with symbols)

[3] The thermophysical properties utilised in the reference simulation are given in Table 7.1.

The maximum width of the FZ for cases V_1 and V_2 agrees well with the experimental data in the web and foot region and a minimum melting of 3 mm is also satisfied, which the reference simulation failed to satisfy, especially in the foot region. Furthermore, for the case V_3, it is observed that the maximum width of the FZ is the largest when compared to the other combinations investigated. Thus, the combination of density values in solid and liquid phases has a considerable influence on the final weld profile.

Dynamic viscosity

Figure 7.13 illustrates the influence of the dynamic viscosity on the maximum width of the FZ. Changing the dynamic viscosity from 5×10^{-3} kg m^{-1} s^{-1} in the reference case to 2.2×10^{-3} kg m^{-1} s^{-1} [18, 105] in V_4 shows no impact on the maximum width of the FZ. Moreover, the minimum melting criteria of 3 mm is not met in the foot region. Thus, the dynamic viscosity values of the liquid steel considered in this study show no influence on the final weld profile.

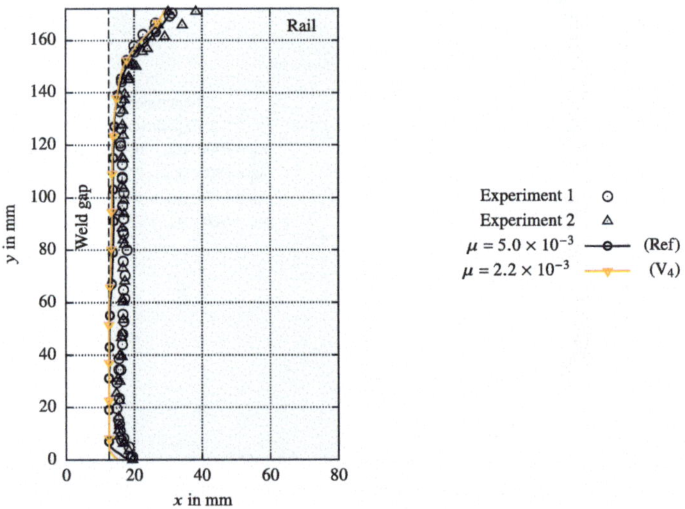

Figure 7.13 Comparison of the maximum width of the FZ on the symmetry plane along the x-axis between: Experiments (symbols) versus reference simulation (black solid line with symbol) with an alternative dynamic viscosity in kg m^{-1} s^{-1} (coloured solid line with symbols)

Specific heat capacity

The numerically obtained maximum width of the FZ for three distinct specific heat capacities of liquid, namely, 732 J kg^{-1} K^{-1} [106] (in V_5), 813.6 J kg^{-1} K^{-1} [25] (in V_6), 950 J kg^{-1} K^{-1} [107] (in V_7), along with the experimental data and reference simulation are shown in Figure 7.14. It is observed that the width of the FZ is narrow when $c_{p,\text{liq}}$ = 732 J kg^{-1} K^{-1} and widens as $c_{p,\text{liq}}$ increases to 950 J kg^{-1} K^{-1}. It is evident that the width of the FZ increases proportionally as the value of $c_{p,\text{liq}}$ increases, indicating that $c_{p,\text{liq}}$ plays a significant role in determining the maximum width of the FZ. Additionally, an extra curve in Figure 7.14 represents the specific heat capacity of the rail steel, $c_{p,\text{sol}}$ = 550 J kg^{-1} K^{-1} [25] (in V_8), as opposed to 640 J kg^{-1} K^{-1} in the reference simulation. Interestingly, in this case, the change in the $c_{p,\text{sol}}$ value has no impact on the maximum width of the FZ.

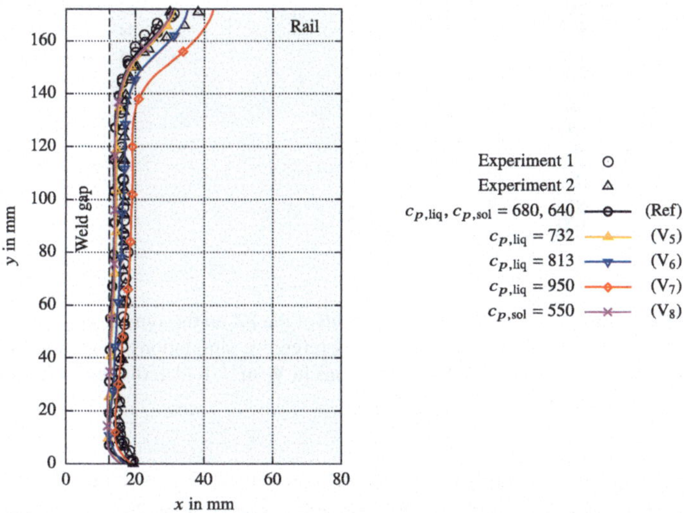

Figure 7.14 Comparison of the maximum width of the FZ on the symmetry plane along the x-axis between: Experiments (symbols) versus reference simulation (black solid line with symbol) with various specific heat capacity values in J kg^{-1} K^{-1} (coloured solid lines with symbols)

Thermal conductivity

Figure 7.15 illustrates the influence of thermal conductivity of the liquid Thermit-steel on the maximum width of the FZ. Changing the thermal conductivity value from 55 W m^{-1} K^{-1} in the reference case to 34 W m^{-1} K^{-1} [108–110] (in V_9) and 45 W m^{-1} K^{-1} (in V_{10}) has shown only a marginal influence on the FZ near the railhead region.

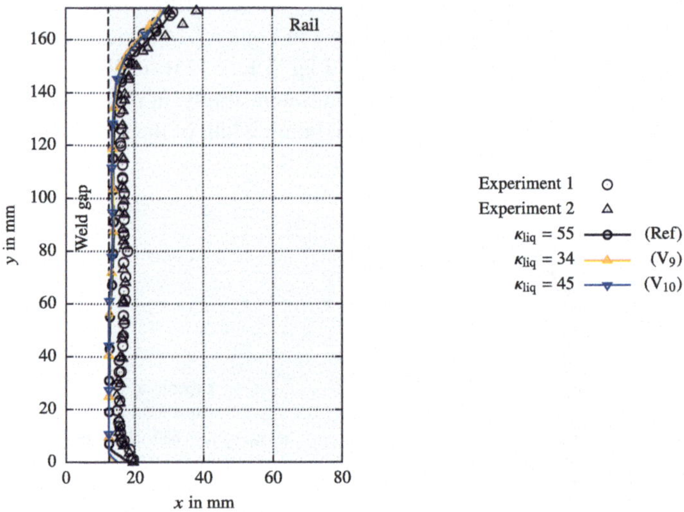

Figure 7.15 Comparison of the maximum width of the FZ on the symmetry plane along the x-axis between: Experiments (symbols) versus reference simulation (black solid line with symbol) with various thermal conductivity values in W m^{-1} K^{-1} (coloured solid lines with symbols)

Table 7.5[4] presents a summary of the influence of various thermophysical properties investigated in this study on the maximum width of the FZ. It is consistently observed that the critical region is located at the rail base ($y < 40$ mm), where the maximum width of the FZ is below the 3 mm limit and is highlighted in grey. In general, the selection of the right set of thermophysical properties is deemed crucial when employing the simplified numerical model for predicting the solidification process during the AT welding of rails.

[4] For the sake of simplicity, the values in the table have been rounded off.

Table 7.5 Comparing the measurements of the maximum width of the FZ (A–Z in Figure 5.11) for all simulated variants

		A	B	C	D	E
		mm	mm	mm	mm	mm
Exp. 1		65	32	34	28	39
Exp. 2		69	34	33	32	39
Ref		62	33	28	25	39
Variant (V)	1	80	43	34	26	39
	2	76	40	32	26	39
	3	101	49	41	30	40
	4	58	32	27	25	30
	5	63	35	39	25	31
	6	70	37	32	26	34
	7	85	43	38	29	40
	8	61	36	27	25	31
	9	55	30	27	25	30
	10	57	30	27	25	30

7.2.6 Rapid Welding Process with Short Preheating (RWSP)

The standard welding procedure employed for the joining and repairing of rails within the railway industry is the RWSP process. This procedural method involves the preheating of the complete casting system with Smartweld jet for an approximate duration of 2.5 min. Following the preheating phase, there is a tapping period of approximately 30 s. During this interval the crucible is positioned above the casting mould assembly and the Thermit-mixture is ignited. In this part of the work, the standard RWSP procedure of the AT welding of rails is numerically investigated and compared with the available experimental data.

Simulation of the preheating stage

A similar approach, as discussed by Chen et al. [13] and Josefson et al. [20], is employed for the numerical determination of the temperature field within the rails during the preheating stage, followed by the tapping period. Herein, the temperature measurements obtained from experiments as discussed in Sect. 5.4, are utilised to calculate the heat flux profile on the rail surface that is in direct contact with the flame from the jet. This heat flux profile is subsequently imposed as a boundary condition on the rail face within the mould. A series of simulations[5] is performed, and the heat flux profile is adjusted (after every simulation) until good acceptance with the experimental data is achieved. Figure 7.16 the comparison of the numerically obtained temporal evolution of the temperature field at different locations (depicted in Figure 5.8) in the rail with the available experimental data. A good agreement with the experiment is observed during the preheating stage. However, during the tapping period, a faster cooling is observed in comparison to the experiment. This discrepancy is attributed to the exclusion of the casting mould in the preheating simulations.

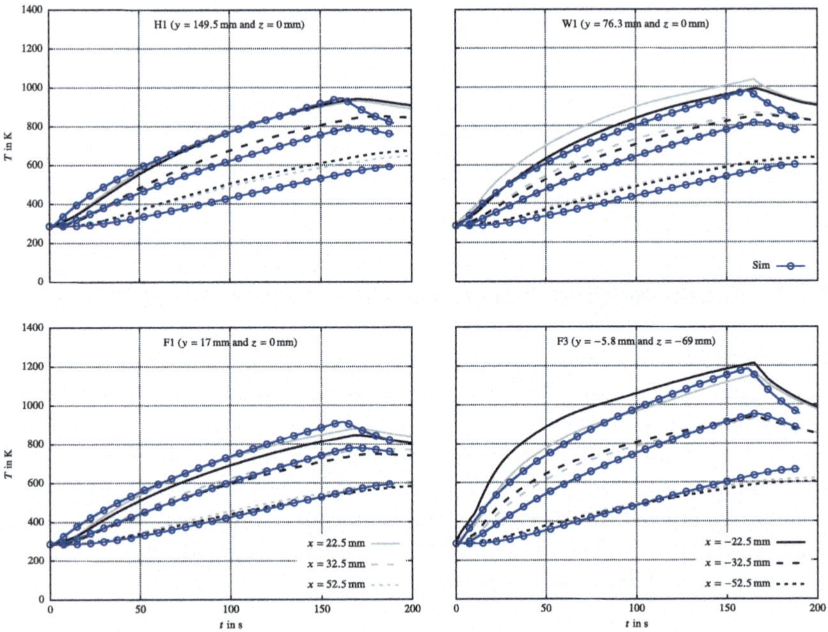

Figure 7.16 Temporal evolution of the temperature field at different locations in the rail during preheating and tapping period: Experiment vs numerical prediction

[5] The simulations are performed using the solver, `laplacianFoam` in OpenFOAM. The solver is capable of solving a transient Laplace equation, for example, thermal diffusion in solids.

Simulation set-up to predict the solidification process

The computational domain, grid resolution, initialisation of the fields (such as phase fraction, volume fraction, velocity), boundary conditions and the numerical schemes remain consistent with those discussed in Sect. 7.1. The only exception is the initialisation of the temperature field (or temperature distribution) in the rail. In this case, the temperature field in the rail is initialised using the temperature distribution obtained from the simulations of the preheating stage, including the tapping period, i.e., at t =190 s, as illustrated in Figure 7.17.

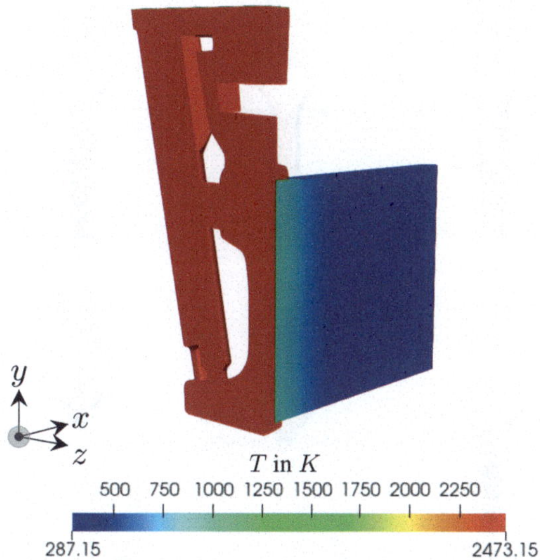

Figure 7.17 The molten metal inside the mould is initialised with the temperature of 2473 K and the temperature distribution in the rail after the preheating stage (including tapping period) at t =190 s

Melting/Solidification

Figure 7.18 illustrates the temporal and spatial evolution of the solid-liquid interface front for the RWSP process. The solidification process is initiated immediately after the preheating stage, with the pouring stage considered instantaneous. At t = 190 s (i.e., after the preheating and tapping stage), the region within the mould is initialised

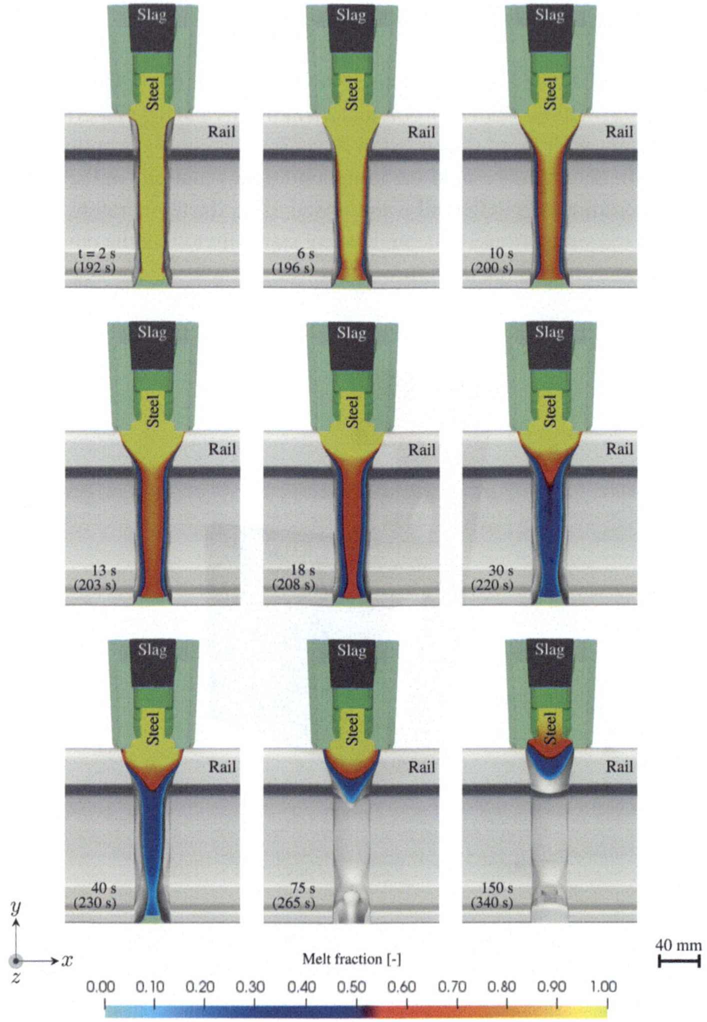

Figure 7.18 Temporal and spatial evolution of the solid-liquid interface front for the initial molten metal temperature of 2473 K with approx 2.5 min of preheating followed by approx 30 s of tapping period

with the molten metal of temperature 2473 K. Figure 7.18 includes two timestamps, with the first indicating the commencement of the solidification simulation and the second (in brackets) denotes the total time, including the preheating stage. As evident from the simulation results, the solid-liquid interface front progresses inwards from the rail end towards the weld centre and upward from the rail base towards the railhead, leading to the formation of a V-shaped solidification pattern.

Comparison with rapid welding process without preheating

It is interesting to compare the solid-liquid interface front at different times between RWSP and RWOP (as depicted in Figure 7.8) processes to illustrate the influence of the preheating stage on the solidification process. Figure 7.19 shows a quantitative analysis of the location of the solid-liquid interface front at different times. Based on Figures 7.8, 7.18 and 7.19 it is evident, that the preheating of rails has a significant impact on the temporal and spatial evolution of the solid-liquid interface front and

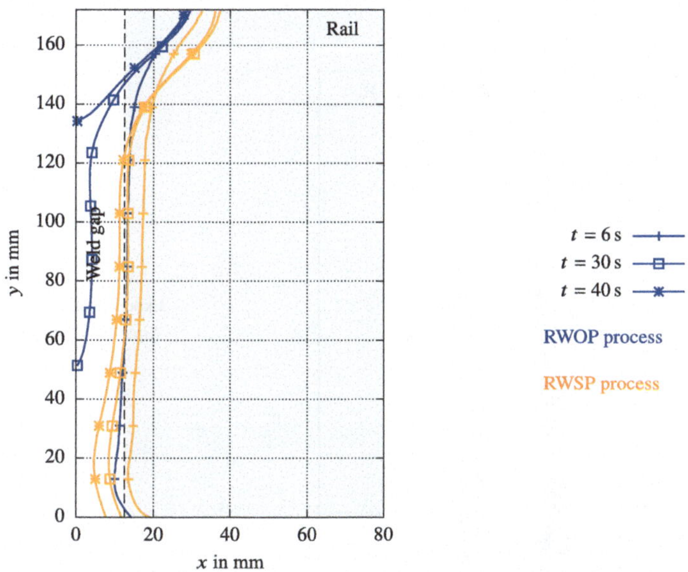

Figure 7.19 Comparison of the location of the solid-liquid interface front on the symmetry plane along the x-axis between RWOP and RWSP processes for the initial molten metal temperature of 2473 K at $t = 6$ s, 30 s, 40 s

on the solidification rate. The width of the solid-liquid interface front from the weld centre evaluated on the symmetry plane along the x-axis is observed to be wider in the RWSP process compared to the RWOP process. At $t = 40$ s the rail base and rail web regions are entirely solidified during the RWOP process, while in the RWSP process, the weld gap remains in the mushy state. Moreover, a slower solidification rate is exhibited by the RWSP process.

Final weld profiles

In Figure 7.20, the experimentally obtained maximum width of the FZ and HAZ is compared with the numerical results evaluated on the symmetry plane along the x-axis. A good agreement is observed between the simulation results and the experiments. It is noted that the maximum width of the FZ in the rail web and rail base region is wider in the experiments than in the simulations. This discrepancy is attributed to consideration of the pouring stage as instantaneous and the assumption of a flat rail surface in the simulations. Conversely, the numerically obtained maximum width of the HAZ over-predicts when compared to the experimental data. This

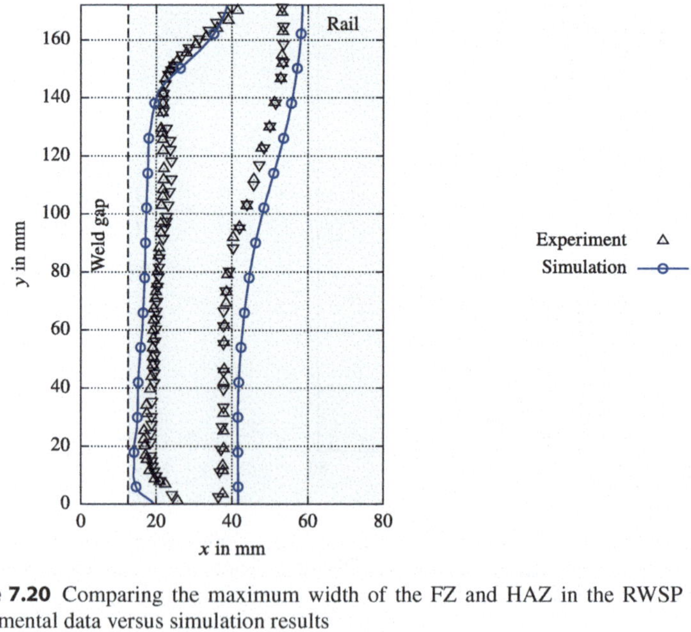

Figure 7.20 Comparing the maximum width of the FZ and HAZ in the RWSP process: Experimental data versus simulation results

disparity arises due to the non-inclusion of the casting mould and non-consideration of heat loss due to radiation in the simulation.

* * *

In this study, a comprehensive investigation into the solidification process during the cooling stage of the AT welding of rails was conducted, employing a laminar flow field assumption. It was assumed that the casting mould was already filled with molten metal and that the rail inside the mould was a flat surface. As a result, the crucible was not included in the simulations, and the domain was reduced to one-quarter of the geometry by taking advantage of the bilateral symmetry of the welding set-up. In the rapid welding process without preheating (RWOP), an investigation into the influence of molten metal temperature on the solidification process revealed dominance of the horizontal direction in the temporal evolution of the solid-liquid interface front for temperatures of 2073 K and 2273 K, resulting in isolated liquid regions in the rail web region. Conversely, a V-shaped temporal evolution was observed for temperatures of 2473 K and 2673 K. Moreover, the significant impact of density and specific heat of Thermit-steel on the maximum width of the FZ was observed.

In the rapid welding process with short preheating (RWSP), evaluation of a heat flux profile from available experimental data was implemented as a boundary condition on the rail face within the mould. Subsequent adjustments to the heat flux profile were made until acceptable agreement with experiments was achieved [13, 20]. Furthermore, the selected temperature distribution in the rail after the end of the preheating stage, followed by a tapping period, was utilised as an initial condition for the solidification simulation. The simulation, conducted for a molten metal temperature of 2473 K, allowed for a comparison of the temporal evolution of the solid-liquid interface front at different times with the RWOP process, showing a significant impact on the solidification process.

Closure 8

Conclusions

In this thesis, a multi-fluid, multiphase, numerical model was developed to investigate the flow field, heat transfer and solidification process during the AT welding of rails. The focus of the study was on understanding the thermal flow field, solid-liquid phase change during the pouring stage, and the temporal evolution of the solid-liquid interface front during the solidification process. The numerical investigations were performed using the FVM-based open-source CFD code OpenFOAM. The multiphase flow solver, `interFoam`, for simulating two incompressible immiscible fluids using the VoF method was modified to incorporate the energy equation and predict the heat transfer within the system. Subsequently, the source code was further extended with the enthalpy-porosity model to predict the solid-liquid phase change process. Additionally, the standard $k - \varepsilon$ turbulence model, which was used to predict the turbulent flow field within the mould during the pouring stage, was also modified to account for the solidification process.

A comprehensive validation for the developed numerical model was discussed, encompassing a diverse range of well-documented cases from the existing literature. Firstly, the numerical model was investigated by selecting a 1D Stefan problem. The numerically obtained location of the solid-liquid interface front and the temperature distribution in the solid were found to be in excellent agreement with the analytical solution. Then, the melting of gallium within a rectangular enclosure was studied, and simulations were performed on both 2D and 3D computational domains. The simulation results showed good agreement with the other numerical results from the literature, and deviations with the reported experimental data were discussed in detail. Finally, the solidification of bismuth metal in the presence of argon gas was examined to unlock (i.e., interface capturing using the VoF method, solid-liquid

R. G. Kewalramani, *Computational Thermo-Fluid Dynamics of Aluminothermic Welding Process*, https://doi.org/10.1007/978-3-658-47535-2_8

phase change using the enthalpy-porosity method) the multi-physics capabilities of the developed model. However, this required further extending the developed model with the Marangoni effect[1]. The numerical results showed excellent agreement with the previously discussed findings in the literature.

Moreover, an in-situ experimental set-up was designed and fabricated to investigate the melting of wax (RT31) within an enclosure in the presence of a gas phase (air). This was conducted to analyse the limitations of the developed model and to gain a better understanding of the discrepancies between experiments and simulations. This validation study demonstrates the accuracy of the developed multiphase model in OpenFOAM.

The data required to validate/verify the numerical results were obtained through a series of experiments conducted within the scope of this thesis. The main challenge was faced in acquiring the shape of the melted rails within the mould immediately after the completion of the pouring stage. Special experiments were conceptualised and conducted at the Technology Innovation Center of the Goldschmidt Holding Group, where a three-piece mould was used instead of a traditional two-piece mould in order to make arrangements to facilitate the draining of the molten metal from the mould immediately after the pouring stage. The photogrammetry technique was adopted to digitalise the experimentally obtained melted rails, and was used to verify the numerical results obtained from the thermal pouring stage simulations, both qualitatively and quantitatively. Additional experiments were performed to obtain the temporal evolution of the temperature field at various locations within the rail for the rapid welding process without preheating (RWOP) of the AT welding of rails. Subsequently, the rail joints were longitudinally sectioned along the rails to evaluate the maximum width of the FZ and HAZ. Although, valuable insights were provided by the experimental results, they were deemed insufficient for conducting a comprehensive uncertainty analysis, and thus the accuracy and reproducibility of the findings reported in this study are called into question.

The thermal pouring (crucible discharge and mould-filling) stage simulations were performed for the RWOP process, i.e., the pre-heating stage is not considered, and the rails were initialised at a constant temperature of 293 K. The temperature of the liquid Thermit-steel within the crucible was initialised at 2473 K. A grid-independent study was conducted and a computation domain of approximately eight million cells was chosen for the investigations reported in this study. The simulations were performed on the HPC-Cluster of the TU BERGAKADEMIE FREIBERG, employing 256 cores. During the mould filling process, the super-heated

[1] The Marangoni effect was considered to discuss this one particular test case, as it was quite challenging to obtain reference/benchmark cases to validate the developed numerical model.

liquid Thermit-steel discharged from the crucible first hits the sand plug and was split into two streams before converging at the railhead-web intersection. The solid rail in its path was melted by super-heated liquid Thermit-steel. Therefore, as the stream of liquid Thermit-steel reaches at the bottom of the mould, its temperature was lower than the initial pouring temperature.

Moreover, the influence of the pouring temperature on the rail profile within the mould and the width of the weld gap post-pouring stage was also investigated. The measuring of the temperature of the liquid Thermit-steel produced after the Thermit-reaction was found to be extremely challenging. The numerically obtained melted rails within the mould were compared with the experimental data. On the one hand, it was evident that the shape of the melted rails within the mould was significantly influenced by the pouring temperature. On the other hand, the numerically obtained shape of the melted rails slightly deviated along the height of the rail (i.e., from railhead to rail base) when compared to the experiments. These deviations can be explained as follows: (a) In the railhead region, air bubbles were introduced within the system due to the low viscosity of the liquid Thermit-steel and the intricate shape of the mould. Although most of the air bubbles were able to escape from the liquid Thermit-steel, a large air bubble was trapped just above the railhead region, leading to poor heat transfer between the liquid Thermit-steel and the railhead. (b) At the railhead-web intersection, the heat transfer was more enhanced in this region. This phenomenon was attributed to non-consideration of the casting mould and sand-plug, and were instead treated as adiabatic boundary conditions. Moreover, it should be noted that, in practice, the rails are slightly elevated to compensate for solidification shrinking during the cooling stage. As a consequent, the weld gap between the rails varied from rail base to railhead. However, in the simulations, a uniform weld gap between the rails was assumed. (c) In the simulations, the melting of the rails below the rail web region was underestimated compared to experiments. This discrepancy was attributed to the draining of the molten metal (i.e., liquid Thermit-steel and slag) from the mould post-pouring stage in the experiment, leading to further melting of the rails in the process.

The simulations of the thermal pouring stage, along with the solid- liquid phase change were computationally expensive. Consequently, assumptions had to be made to simplify both the physics involved during the AT welding process and the computational domain. This allowed the investigation of the influence of various parameters on the spatial and temporal evolution of the solid-liquid interface front, FZ and HAZ in a shorter period. The problem at hand was simplified by assuming that the casting mould was pre-filled with the molten metal, and the rail end surface inside the mould was a flat surface, indicating that the rail end was not melted off during the pouring stage. This assumption was quite common in the literature to avoid the computational resources needed to simulate the turbulent thermal pouring stage, led to skipping the pouring stage altogether. Additionally, due to the bilateral symmetry of the geometry, only one quarter of the computational domain was considered. This has reduced the number of cells by a factor of 5 and the number of cores required to predict final weld profiles by a factor of 16.

Measuring the thermophysical properties of the Thermit-steel above the liquidus temperature posed a challenging task. Therefore, literature involving pure iron (or steel composition comparable to that of Thermit-steel) was relied upon to determine the thermophysical properties used in the study.

The spatial and temporal evolution of the solid-liquid interface front was discussed in detail for the initial molten metal temperature of 2473 K. A V-shaped solidification process was observed, wherein the solid-liquid interface front progressed inwards from the rail end towards the weld centre and upwards from the rail base towards the railhead. This process is essential for the AT welding process to prevent blocking of the liquid feeding path, thereby avoiding the formation of an isolated melt region within the FZ, which could result in shrinkage defects. Moreover, it was observed that the flow field inside the mould was initially dominated by thermal convection, inducing a circular flow within the mould.

The impact of the initial molten metal temperature on the temporal and spatial evolution of the solid-liquid interface front and the final weld profile was also examined. It was noted that lower temperatures (i.e., 2073 K and 2273 K) posed a risk of forming an isolated liquid region, leading to shrinkage defects or centreline fracture. Furthermore,the minimum melting criteria of the parent rail for the joining process, which is 3 mm as per the European standard, were not met by the rail-joint. Additionally, the influence of different thermophysical properties from the literature on the maximum width of the FZ was investigated. The results indicate that for the values examined in this study, a significant influence was exerted by the combination of rail (solid) and Thermit-steel (liquid) density, as well as the specific heat capacity of liquid. Thermal conductivity, on the other hand had a marginal effect in the railhead region, while dynamic viscosity had no impact on the maximum width of the FZ.

Efforts were made to predict the solidification process for the rapid welding process with preheating (RWSP). The heat flux profile was evaluated from available experimental data to predict the temperature distribution in the rail during the preheating stage. This was then implemented as a boundary condition on the rail face within the mould. Transient heat condition simulations were performed using the solver `laplacianFoam` for upto 2.5 min of preheating, followed by 30 s of tapping. A series of simulations were performed, and the heat flux profile was adjusted every time until good acceptance with the available experimental data was observed.

The numerically obtained temperature distribution in the rail was then imposed as an initial condition to predict the spatial and temporal evolution of the solid-liquid interface front and the maximum width of the FZ and HAZ. Furthermore, the simulation results were compared with those of the RWOP process, and the impact of the preheating stage was demonstrated. The width of the solid-liquid interface front was more comprehensive compared to that of the RWOP process. Moreover, the solid-liquid interface front progressed slower than the RWOP process.

* * *

The numerical model developed in the scope of this thesis, provides valuable insights into the flow field, heat transfer and solidification process during the pouring and cooling stages of the AT welding of rails. The numerical model can serve as a reliable tool to optimise process by conducting parametric studies, which may include improving, redesigning or downsizing the casting system, improving the solidification process, and so on. Moreover, the developed numerical model can be applied to various other applications in the foundry industry, including but not limited to gravity casting.

Future work

Although the developed numerical model provides insightful information into the pouring and cooling stages of the process, there is still room for further improvements. Therefore, the following recommendations are proposed for future investigations:

- In this thesis, the simulation of the pouring stage involved a two-fluid system, specifically Thermit-steel and air. However, the actual AT welding process involves a three-fluid system consisting of Thermit-steel, slag and air. Therefore, the existing model has to be extended to include the third fluid (i.e., slag), to investigate the dynamics of both Thermit-steel and slag during the pouring stage, which will help in enhancing the overall understanding of the pouring stage.
- The casting mould and the sand plug were not explicitly included in the simulations, instead, they were implemented as adiabatic boundary condition. To improve the model, it is suggested to include the conjugate heat transfer (CHT) between the molten metal and the casting mould.
- The current numerical study utilises constant thermophysical properties for each phase i.e., for solid (rail) and liquid (Thermit-steel). To further enhance the de-

veloped model, it is recommended to incorporate temperature-dependent thermophysical properties.
- The simulation of the preheating stage during the AT welding process is quite complex and beyond the scope of this thesis. Nevertheless, a simple heat conduction model was employed to understand the influence of the preheating stage on the solidification process. A heat flux boundary condition was imposed on the rail face within the mould to predict the temperature distribution in the rail. Therefore, to achieve more accurate results, it is recommended to conduct full-fledged CFD simulations to predict the temperature distribution within the casting system during the preheating stage of the AT welding process.

* * *

The proposed recommendations will pave the way for significant advancements in the numerical modelling of the AT welding process for joining rails in OpenFOAM. Furthermore, the integration of these improvements will lay the groundwork for the implementation of a Digital Twin framework. This will serve as a virtual replica for continuous monitoring, analysis and optimisation, ultimately leading to a more efficient and reliable rail joining process.

List of Publications

The findings reported in this study have been published in various international journals and peer-reviewed conferences. Additionally, some of the results have been presented in the form of informative posters and engaging presentations:

Journals

- **R. G. Kewalramani**, I. Riehl, J. Hantusch, T. Fieback: Numerical investigation of the cooling stage during aluminothermic welding of rails: Rapid welding without preheating. *Thermal Science and Engineering Progress (TSEP)*, Vol. 37 (2023), 101610.
- **R. G. Kewalramani**, S. Pose, I. Riehl, T. Fieback: Solid-Liquid Phase Change in the Presence of Gas Phase: Numerical Modeling and Validation. *Journal of Fluid Flow, Heat and Mass Transfer (JFFHMT)*, Vol. 7 (2020), pp. 1-11.

Peer-reviewed Conferences

- **R. G. Kewalramani**, I. Riehl, J. Hantusch, T. Fieback: A CFD Model to simulate Thermal Mould Filling and Solid-Liquid Phase Change during Aluminothermic Welding of Rails. *Proceedings of the 14th International Conference on Computational Heat and Mass Transfer (ICCHMT2023)*, 04. - 08. Sep. 2023, Düsseldorf, Germany. (**accepted**)
- **R. G. Kewalramani**, I. Riehl, J. Hantusch, J. Keichel, M. Niederkrueger, T. Fieback: Numerical Modelling and Simulation of the Aluminothermic Welding of Rails: Heat Transfer and Solid-Liquid Phase Change. *Proceedings of the 7th World Congress on Mechanical, Chemical and Material Engineering (MCM'21)*, 2. - 4. Aug. 2021, Prague, Czech Republic [Virtual Conference].

R. G. Kewalramani, *Computational Thermo-Fluid Dynamics of Aluminothermic Welding Process*, https://doi.org/10.1007/978-3-658-47535-2

- **R. G. Kewalramani**, S. Pose, I. Riehl, T. Fieback: Investigation of the Solid-Liquid Phase Change in the Presence of Gas Phase: Numerical Modeling and Validation. *Proceedings of the 5th World Congress on Mechanical, Chemical and Material Engineering (MCM'19)*, 15. - 17. Aug. 2019, Lisbon, Portugal.

Conferences (Posters/Presentations)

- **R. G. Kewalramani**, S. Pose, I. Riehl, J. Hantusch, T. Fieback: Aluminothermic Welding Process: 3D CFD Simulation of Solid-Liquid Phase Change and Experimental Validation using Photogrammetry. *18th Multiphase Flow Conference & Short Course*, 08. - 12. Nov. 2021, Dresden, Germany [Virtual Conference].
- **R. G. Kewalramani**, I. Riehl, J. Hantusch, T. Fieback: Multiphase CFD Simulation of Heat Transfer and Solidification During Thermite Welding Process. *14th International Conference on Computational Fluid Dynamics in the Oil & Gas, Metallurgical and Process Industries*, 12. - 14. Oct. 2020, Trondheim, Norway [Virtual conference].
- **R. G. Kewalramani**, I. Riehl, T. Fieback: Modellierung und Simulation des Erstarrungsvorganges beim Thermit-Schweißen von Schienen. *Thermodynamik-Kolloquium* 2018, 26. - 28. Sep. 2018, Kassel, Germany.
- **R. G. Kewalramani**, I. Riehl, T. Fieback: CFD-Simulation des Thermit-Schweißverfahrens - Abkühlung und Erstarrung der Schmelze. *Jahrestreffen der ProcessNet-Fachgruppen Mehrphasenströmungen, Wärme und Stoffübertragung, CFD, HTT, AuW, KRI und PMT*, 6. - 9. Mar. 2018, Bremen, Germany.
- **R. G. Kewalramani**, I. Riehl, T. Fieback: Numerische Untersuchung der Wärmeübertragung und Erstarrung der Schmelze im Thermit-Schweißverfahren mittels OpenFOAM. *Thermodynamik-Kolloquium* 2017, 27. - 29. Sep. 2017, Dresden, Germany.

Bibliography

1. United Nations Framework Convention on Climate Change. *The Paris Agreement*. 2015. URL: https://treaties.un.org/Pages/ViewDetails.aspx?src=TREATY&mtdsg_no=XXVII-7-d&chapter=27&clang=_en.
2. "Regulations—European Climate Law". In: *Official Journal of he European Union* 64 (2021).
3. Federal Ministry for the Environment, Nature Conservation, Nuclear Safety and Consumer Protection, ed. *Climate Action Plan 2050—Principles and goals of the German government's climate policy*. Brochure—No. 6037. 2016. URL: https://www.bmuv.de/en/publication/climate-action-plan-2050-en.
4. *Smart and affordable rail services in the EU: a socio-economic and environmental study for High-Speed in 2030 and 2050*. EU-Rail. 2023.
5. Allianz pro Schiene. URL: https://www.allianz-pro-schiene.de/themen/infrastruktur/elektrifizierung-bahn/ (visited on Mar. 8, 2024).
6. Paul Hockenos. *Rail cargo emissions in Germany*. 2018. URL: https://www.cleanenergywire.org/factsheets/rail-cargo-emissions-germany (visited on Mar. 8, 2024).
7. A. Enshaeian and P. Rizzo. "Stability of continuous welded rails: A state-of-the-art review of structural modeling and nondestructive evaluation". In: *Proceedings of the Institution of Mechanical Engineers, Part F: Journal of Rail and Rapid Transit* 235.10 (2021), pp. 1291–1311.
8. Bijendra Jha. "Thermite welding of rail steel". PhD thesis. Illinois Institute of Technology, 1989.
9. Nataliia Tuchkova. "Prozessanalyse und simulationstechnische Optimierung des aluminothermischen Schweißens von Schienen". German. PhD thesis. Otto-von-Guericke-Universität Magdeburg, 2011.
10. J. Wegrzyn and M. Maxur. "A thermit mixture for rail track welding". In: *Welding International* 6 (1 1992), pp. 5–8.
11. DIN EN 13674-1. *Railway applications—Track—Rail—Part 1: Vignole railway rails 46 kg/m and above*. July 2017.
12. Goldschmidt Holding GmbH. *Smartweld Jet*. Goldschmidt Holding GmbH. URL: https://www.goldschmidt.com/en/product/smartweld-jet/ (visited on Mar. 8, 2024).
13. Y. Chen, F. V. Lawrence, and C. P. L. Barkan. *A heat transfer modeling study of rail thermite welding*. Tech. rep. University of Illinois at Urbana Champaign, 2004.

R. G. Kewalramani, *Computational Thermo-Fluid Dynamics of Aluminothermic Welding Process*, https://doi.org/10.1007/978-3-658-47535-2

14. Y. Chen, F. V. Lawrence, C. P. L. Barkan, and J. A. Dantzig. "Heat transfer modelling of rail thermite welding". In: *Proceedings of the Institution of Mechanical Engineers, Part F: Journal of Rail and Rapid Transit* 220.3 (2006), pp. 207–217.
15. Y. Chen, F. V. Lawrence, C. P. L Barkan, and J. A. Dantzig. "Weld defect formation in rail thermite welds". In: *Proceedings of the Institution of Mechanical Engineers, Part F: Journal of Rail and Rapid Transit* 220.4 (2006), pp. 373–384.
16. S. Weiss, I. Riehl, J. Hantusch, and U. Gross. "Numerical investigation on the crucible discharge of steel and slag during the aluminothermic welding process". In: *Archives of Metallurgy and Materials* 63.1 (2018), pp. 173–180.
17. S. Weiss, I. Riehl, J. Hantusch, and U. Gross. "Contact angle measurements of steel on waterglass-bonded Al2O3-SiO2 composite materials". In: *High Temperatures – High Pressures* 47 (6 2018), pp. 525–537.
18. Sebastian Weiss. "Early stages of the aluminothermic process: Insights into separation and mould Filling". PhD thesis. TU Bergakademie Freiberg, 2019.
19. B. L. Josefson and R. Bisschop. *Calculated microstructure, residual deformations and stress fields for the aluminothermic welding process serving weld quality optimisation*. Tech. rep. Chalmers University of Technology, 2018.
20. B. L. Josefson, R. Bisschop, M. Messaadi, and J. Hantusch. "Residual stresses in thermite welded rails: significance of additional forging". In: *Welding in the world* 64 (2020), pp. 1195–1212.
21. Jan Hantusch. *Report with process improvements, equipment design and performance specification*. Tech. rep. Goldschmidt Holding GmbH, 2017.
22. S. Manzke, I. Riehl, T. Fieback, and U. Gross. "A reduced numerical model for the thermit rail welding process". In: *Heat Transfer Engineering* 39 (13–14 2018), pp. 1296–1307.
23. C. R. Swaminathan and V. R. Voller. "A general enthalpy method for modeling solidificatoin process". In: *Metallurgical Transactions B* 23 (1992), pp. 651–664.
24. Anja Miehe. "Numerical investigation of horizontal twin-roll casting of the Magnesium alloy AZ31". PhD thesis. TU Bergakademie Freiberg, 2014.
25. Sebastian Manzke. "Reduzierte thermische Modelle für das gesamte Thermit-Schweißverfahren". German. PhD thesis. TU Bergakademie Freiberg, 2021.
26. I. Riehl, S. Manzke, A. Jünger, R. G. Kewalramani, J. R. Schwarz, and U. Groß. *Entwicklung eines energieeffizienten und modularen Heizsystems zur thermischen Vorbehandlung von Gießsystemen f ü r das aluminothermische Schweißen von Schienen (MOTEVAS). Simulation und Optimierung des AT-Schweißverfahrens*. German. Tech. rep. TU Berg-akademie Freiberg, 2020.
27. M. S. Galay and A. S. Ilinykh. "Improving the technology of aluminothermic rail welding based on software simulation". In: *Journal of Physics: Conference Series* 1967.012063 (2021).
28. Y. Liu, K. S. Tsang, N. A. Subramaniam, and J. J. L. Pang. "Structural fatigue investigation of thermite welded rail joints considering weld-induced residual stress and stress relaxation by cyclic load". In: *Engineering Structures* 235 (2021), p. 112033.
29. Eduardo De Abreu Lima. "Modelo Numérico Termo-Elastoplástico do Processo de Soldagem Aluminotérmica em Trilhos Ferroviários". Portuguese. PhD thesis. Universidade Estadual De Campinas, 2023.
30. Petr A. Nikrityuk. *Computational Thermo-Fluid Dynamics: In Materials Science and Engineering*. WILEY-VCH, 2011.

31. C. W. Hirt and B. D. Nichols. "Volume of Fluid (VOF) method for the dynamics of free boundaries". In: *Journal of Computational Physics* 39 (1981), pp. 201–225.
32. Onno Ubbink. "Numerical prediction of two fluid systems with sharp interfaces". PhD thesis. Imperial College, University of London, 1997.
33. O. Ubbink and R. I. Issa. "A method for capturing sharp fluid interfaces on arbitrary meshes". In: *Journal of Computational Physics* (1999).
34. S. Muzaferija, M. Perić, P. Sames, and T. Schelin. "A two-fluid navier-stokes solver to simulate water entry". In: *Twenty-Second Symposium on hydrodynamics Commission on Physical Sciences, Mathematics and Applications*. 1988, pp. 638–651.
35. David L. Youngs. "Time-dependent multi-material flow with large fluid distortion". In: *Numerical Methods in Fluid Dynamics*. Ed. by K. W. Morton and M. J. Baines. Academic Press, 1982, pp. 273–285.
36. W. F. Noh and P. Woodward. "SLIC (Simple Line Interface Calculation)". In: *Fifth International Conference on Numerical Methods in Fluid Dynamics*. 1976.
37. J. Klostermann, K. Schaake, and R. Schwarze. "Numerical simulation of a single rising bubble by VOF with surface compression". In: *International Journal for numerical methods in fluids* 71 (2013), pp. 960–982.
38. V. Dyadechko and M. Shashkov. *Moment-of-fluid interface reconstruction*. Tech. rep. Los Alamos National Laboratory, 2005.
39. M. Sussman and E. G. Puckett. "A coupled level set and volume-of-fluid method for computing 3D and axisymmetric incompressible two-phase flows". In: *Journal of Computational Physics* 162 (2000), pp. 301–337.
40. M. Skarysz, A. Garmory, and M. Dianat. "An iterative interface reconstruction method for PLIC in general convex grids as part of a coupled level set volume of fluid solver". In: *Journal of Computational Physics* 368 (2018), pp. 254–276.
41. Henrik Rusche. "Computational Fluid Dynamics of dispersed two-phase flows at high phase fractions". PhD thesis. Imperial College of Science, Technology and Medicine, 2002.
42. J. U. Brackbill, D. B. Kothe, and C. Zemach. "A Continuum method for modeling surface-tension". In: *Journal of Computational Physics* 100 (1992), pp. 335–354.
43. A. D. Brent, V. R. Voller, and K. J. Reid. "Enthalpy-porosity technique for modeling convection-diffusion phase change: Application to the melting of a pure metal". In: *Numerical Heat Transfer* 13 (1988), pp. 297–318.
44. V. R. Voller and C. Prakash. "A fixed grid numerical modelling methodology for convection-diffusion mushy region phase-change problmes". In: *Int. J. Heat Mass Transfer* 30 (8) (1987), pp. 1709–1719.
45. Joseph Boussinesq. *Th é orie de l' é coulement tourbillonnant et tumultueux des liquides dans les lits rectilignes a grande section*. Gauthier-Villars et fils, 1897.
46. J. H. Ferziger and M. Perić. *Computational Methods for Fluid Dynamics, 3rd Edition*. Springer-Verlag Berlin Heidelberg, 2002.
47. P. C. Carman. "Fluid flow through granular beds". In: *Chemical Engineering Research and Design* 75 (1997), S32–S48.
48. F. Moukalled, L. Mangani, and M. Darwish. *The Finite Volume Method in Computational Fluid Dynamics: An Advanced Introduction with OpenFOAM and Matlab*. Springer, 2016.
49. B. E. Launder and D. B. Spalding. "The numerical computation of turbulent flows". In: *Computer Methods in Applied Mechanics and Engineering* 3 (1974), pp. 269–289.

50. Claudia Pfeiler. "Modeling of turbulent particle/gas dispersion in the mold region and particle entrapment into the solid shell of a steel continuous caster". PhD thesis. Montanuniversität Leoben, 2008.
51. A. Vakhrushev, A. Ludwig, M. Wu, Y. Tang, G. Hackl, and G. Nitzl. "Coupling the turbulent flow with the solidification processes in OpenFOAM". In: *5th Open Source CFD International Conference*. 2011.
52. H. K. Versteeg and W. Malalasekera. *An Introduction to Computational Fluid Dynamics: The Finite Volume Method, Second Edition*. Harlow, England: Pearson Education Limited, 2007.
53. Hrvoje Jasak. "Error analysis and estimation for the Finite Volume Method with applications to fluid flows". PhD thesis. Imperial College of Science, Technology and Medicine, 1996.
54. Suhas V. Patankar. *Numerical Heat Transfer and Fluid Flow*. Hemisphere Publishing Corporation, 1980.
55. Bram van Leer. "Towards the ultimate conservative difference scheme. II. Monotonicity and conservation combined in a second-order scheme". In: *Journal of Computational Physics* 14 (4 1974), pp. 361–370.
56. B. P. Leonard. "A stable and accurate convective modelling procedure based on quadratic upstream interpolation". In: *Computer Methods in Applied Mechanics and Engineering* 19 (1 1979), pp. 59–98.
57. Milovan Perić. "A Finite Volume Method for the prediction of three-dimensional fluid flow in complex ducts". PhD thesis. Imperial College, University of London, 1985.
58. G. D. Raithby and K.E. Torrance. "Upstream-weighted differencing schemes and their application to elliptic problems involving fluid flow". In: *Computational Fluids* 2 (1974), pp. 191–206.
59. Santiago Márquez Damián. "An extended mixture model for the simultaneous treatment of short and long scale interfaces". PhD thesis. UNIVERSIDAD NACIONAL DEL LITORAL, 2013.
60. S. S. Deshpande, L. Anumolu, and M. F. Trujillo. "Evaluating the performance of the two-phasae flow solver interFoam". In: *Computational Science & Discovery* 5 (2012).
61. V. R. Voller and C. R. Swaminathan. "General source-based method for solidfication phase change". In: *Numerical Heat Transfer, Part B* 19 (1991), pp. 175–189.
62. C. R. Swaminathan and V. R. Voller. "On the enthalpy method". In: *International Journal of Numerical Methods for Heat and Fluid Flow* 3 (1993), pp. 233–244.
63. Yousef Saad. *Iterative Methods for Sparse Linear Systems, Second Edition*. United States of America: Society for Industrial and Applied Mathematics, 2003.
64. C. M. Rhie and W. L. Chow. "Numerical study of the turbulent flow past an airfoil with trailing edge separation". In: *AIAA Journal* 21 (1983), pp. 1525–1532.
65. S. V. Patankar and D. B. Spalding. "A calculation procedure for heat, mass and momentum transfer in three-dimensional parabolic flows". In: *International Journal of Heat and Mass Transfer* 15 (1972), pp. 1787–1806.
66. R. I. Issa. "Solution of the implicit discretized fluid flow equations by operator splitting". In: *Journal of Computational Physics* 62 (1986), pp. 40–65.
67. R. Courant, K. O. Friedrichs, and H. Lewy. "Über die partiellen Differentialgleichungen der mathematischen Physik". German. In: *Mathematische Annalen* 100 (1 1928), pp. 32–74.

68. E. Berberović, S. Jakirlić N. P. van Hinsberg, I. V. Roisman, and C. Tropea. "Drop impact onto a liquid layer of finite thickness: Dynamics of the cavity evolution". In: *Physical Review E* 79 (2009).
69. H. G. Weller, G. Tabor, H. Jasak, and C. Fureby. "A tensorial approach to computational continuum mechanics using object-oriented techniques". In: *Computers in Physics* 12.620–631 (1998).
70. H. Jasak, A. Jemcov, and Ž. Tuković. "OpenFOAM: A C++ library for complex physics simulations". In: *International Workship on Coupled Methods in Numerical Dynamics* (2007).
71. R. G. Kewalramani, S. Pose, I. Riehl, and T. Fieback. "Solid-Liquid Phase Change in the Presence of Gas Phase: Numerical Modeling and Validation". In: *Journal of Fluid Flow, Heat and Mass Transfer* 7 (2020), pp. 1–11.
72. J. A. Dantzig and M. Rappaz. *Solidification*. 1st. EPFL Press (A Swiss academic publisher distributed by CRC Press), 2009.
73. C. Gau and R. Viskanta. "Melting and solidification of a pure metal on a vertical wall". In: *Journal of Heat Transfer* 108 (1986), pp. 174–181.
74. R. R. Kasibhatla and D. Brüggemann. "Smoothed iterative enthalpy approach for solid-liquid phase change". In: *International Journal of Thermal Sciences* 152 (2020).
75. A. König-Haagen, E. Franquet, E. Pernot, and D. Brüggemann. "A comprehensive benchmark of fixed-grid methods for the modeling of melting". In: *International Journal of Thermal Sciences* 118 (2017), pp. 69–103.
76. V. Kumar, F. Durst, and S. Ray. "Modeling moving-boundary problems of solidification and melting adopting an arbitrary lagrangian-eulerian approach". In: *Numerical Heat Transfer* 49 (2006), pp. 299–331.
77. F. Rösler and D. Brüggemann. "Shell-and-tube type latent heat thermal energy storage: numerical analysis and comparison with experiments". In: *Heat and Mass Transfer* 47 (2011), pp. 1027–1033.
78. Zaki Saptari Saldi. "Marangoni driven free surface flows in liquid weld pools". PhD thesis. Delft University of Technology, 2012.
79. K. Wittig and P. A. Nikrityuk. "Three-dimensionality of fluid flow in the benchmark experiment for a pure metal melting on a vertical wall". In: *IOP Conf. Series: Materials Science and Engineering* 27 (2011).
80. Jonathan A. Dantzig. "Modelling liquid-solid phase changes with melt convection". In: *International Journal for Numerical Methods in Engineering* 28 (1989), pp. 1769–1785.
81. F. Stella and M. Giangi. "Melting of a pure metal on a vertical wall: Numerical simulation". In: *Numerical Heat Transfer, Part A: Applications* 38 (2 2000), pp. 193–208.
82. N.e Hannoun, V. Alexiades, and T. Z. Mai. "Resolving the controversy over tin and gallium melting in a rectangular cavity heated from the side". In: *Numerical Heat Transfer, Part B* 44 (2003), pp. 253–276.
83. C. Gau and R. Viskanta. "Melting and solidification of a metal system in a rectangular cavity". In: *International Journal of Heat and Mass Transfer* 27.1 (1984), pp. 113–123.
84. L. H. Tan, S. S. Leong, T. J. Barber, and E. Leonardi. "Simulation of solidification with marangoni effects using a multiphase approach". In: 2005.
85. T. Yamamoto, Y. Okano, and S. Dost. "Validation of the S-CLSVOF method with the density-scaled balanced continuum surface force model in multiphase systems coupled with thermocapillary flows". In: *International Journal for Numerical Methods in Fluids* 83 (2017), pp. 223–244.

86. R. R. Kasibhatla, A. König-Haagen, and D. Brüggemann. "Numerical modelling of wetting phenomena during melting of PCM". In: *Procedia Engineering* 157 (2016), pp. 139–147.
87. Fabian Rösler. "Modellierung und Simulation der Phasenwechselvorgänge in makroverkapselten latenten thermischen Speichern". German. PhD thesis. Universität Bayreuth, 2014.
88. Y. Wang, A. Amiri, and K. Vafai. "An experimental investigation of the melting process in a rectangular enclosure". In: *International Journal of Heat and Mass Transfer* 42 (1999), pp. 3659–3672.
89. *Betriebsanleitung: LAUDA Ecoline Staredition RE 2XX*. Version 01/2018 f. LAUDA DR. R. WOBSER GMBH & CO. KG. LAUDA DR. R. WOBSER GMBH & CO. KG Pfarrstraße 41/43 97922 Lauda-Königshofen Deutschland, 2018.
90. Rubitherm Technologies GmbH. URL: https://www.rubitherm.eu/en/index.php/productcategory/organische-pcm-rt (visited on Apr. 11, 2022).
91. H. Shmueli, G. Ziskind, and R. Letan. "Melting in a vertical cylindrical tube: Numerical investigation and comparison with experiments". In: *International Journal of Heat and Mass Transfer* 53 (2010), pp. 4082–4091.
92. *Technische Information: Plexiglas® GS/XT*. Evonik Industries AG. Evonik Industries AG Acrylic Polymers Kirschenallee, 64293 Darmstadt, Deutschland, 2011.
93. Thomas Reinicke. *Skilled in the hands-on implementation of AT Welding. Project Technician*. GOLDSCHMIDT HOLDING GMBH Technology Innovation Center Am Schenkberg 20, 04349 Leipzig, Germany, 2019.
94. R. G. Kewalramani, I. Riehl, T. Reinicke, and J. Hantusch. *Planning and overseeing experiments related to the AT Welding process as documented in this study*. GOLDSCHMIDT HOLDING GMBH Technology Innovation Center Am Schenkberg 20, 04349 Leipzig, Germany, 2019.
95. *Agisoft Metashape Professional*. Version 1.7.5. 2021. URL: https://www.agisoft.com/downloads/installer/.
96. Icons by svgrepo.com. URL: https://www.svgrepo.com/ (visited on Apr. 11, 2022).
97. R. G. Kewalramani, I. Riehl, J. Hantusch, and T. Fieback. "Numerical investigation of the cooling stage during aluminothermic welding of rails: Rapid welding process without preheating". In: *Thermal Science and Engineering Progress* 37 (2023), p. 101610.
98. Ankit Rohatgi. *Webplotdigitizer: V 4.2*. URL: https://automeris.io/WebPlotDigitizer.
99. R. G. Kewalramani, I. Riehl, J. Hantusch, and T. Fieback. "A CFD model to simulate thermal mould-filling and solid-liquid phase change during aluminothermic welding of rails". In: *Proceedings of the 14. International Conference on Computational Heat and Mass Transfer (**accepted**)*. Springer Nature, 2024.
100. VDI e. V., ed. *VDI-Wärmeatlas*. 11th ed. Springer Vieweg Berlin, Heidelberg, 2013.
101. Ronald Oertel. "Numerische Untersuchungen zur Entwicklung der Phasengrenze beim Aufschmelzen und Erstarren von Stahl". German. MA thesis. TU Bergakademie Freiberg, 2013.
102. Brian Gordon Thomas. "Investigation of Panel Crack Formation in Steel Ingots using Mathematical and Physical Models". PhD thesis. University of British Columbia, 1985.
103. F. Pellegrini and J. Roman. "Scotch: A software package for static mapping by dual recursive bipartitioning of process and architecture graphs". In: *High-Perormance Computing and networking*. Springer Berlin Heidelberg, 1996, pp. 493–498.

104. DIN EN 14730-I. *Railway applications—Track—Aluminothermic welding of rails—Part 1: Approval of welding process*. 2017.
105. M. J. Assael, K. Kakosimos, R. M. Banish, J. Brillo, I. Egry, R. Brooks, P. N. Quested, K. C. Mills, A. Nagashima, Y. Sato, and W. A. Wakeham. "Reference data for the density and viscosity of liquid aluminum and liquid iron". In: *Journal of Physical and chemical Reference Data* 35 (1 2006), pp. 285–300.
106. G. Pottlacher, H. Jäger, and T. Neger. "Thermophysical measurements on liquid iron and nickel". In: *High Temperatures—High Pressures* 19 (1987), pp. 19–27.
107. K. C. Mills. *Recommended values of thermophysical properties for selected commercial alloys*. Woodhead Publishing, 2002.
108. P. Oksman, S. Yu, H. Kytönen, and S. Louhenkilpi. "The effective thermal conductivity method in continuous casting of steel". In: *Acta Polytechnica Hungarica* 11.9 (2014).
109. I. Staniewski and W. Derda. "Application of the CFD computing technique to numerical modeling of continuous casting of steel". In: *Achives of Metallurgy and Materials* 50 (4 2005).
110. Z. Zhang, M. Wu, H. Zhang, S. Hahn, F. Wimmer, A. Ludwig, and A. Kharicha. "Modeling of the as-cast structure and macrosegregation in the continuous casting of a steel billet: Effect of M-EMS". In: *Journal of Materials Processing Tech.* 301 (2022).

The manufacturer's authorised representative in the EU is Springer Nature Customer Service Centre GmbH, Europaplatz 3, 69115 Heidelberg, Germany. If you have any concerns regarding our products, please contact ProductSafety@springernature.com

Printed and bound by CPI Group (UK) Ltd, Croydon, CR0 4YY
31/07/2026
02182786-0001